...REPAIRS AND INSPECTION

Proceedings of a session sponsored by
the Structural Division of the
American Society of Civil Engineers
in conjunction with the ASCE Convention
in Atlantic City, New Jersey

April 29, 1987

Edited by Mahendra J. Shah

Published by the
American Society of Civil Engineers
345 East 47th Street
New York, New York 10017-2398

ABSTRACT

Rehabilitation of a structure in general requires considerations of unique methods of inspection and repair. Since the structures have to remain in service while the repairs are done, the task of rehabilitation becomes especially challenging. The papers in this volume examine how special methods of inspection and repair were developed for bridges, viaduct, and marine structures to minimize the cost of rehabilitation while maintaining the service requirements of these structures.

Library of Congress Cataloging-in-Publication Data

Infrastructure—repairs and inspection.

Includes indexes.
1. Public works—Maintenance and repair—Congresses. 2. Public works—New York (N.Y.)—Maintenance and repair—Congresses. I. Shah, Mahendra J. II. American Society of Civil Engineers. Structural Division. III. ASCE National Convention (1987 : Atlantic City, N.J.)
TA5.I425 1987 624'.028'8 87-1267
ISBN 0-87262-584-2

Library of Congress Catalog Card No: 86-1267
ISBN 0-87262-584-2
Manufactured in the United States of America.

PREFACE

One does not have to go far to see that our infrastructure needs repair. Roads and bridges constructed during the 1960s and earlier, including many of our landmark structures, have been neglected to the extent that many of them are unsafe for the increased service they are challenged to provide. Therefore, a session on Infrastructure Repairs and Inspection was organized by the Structural Division of the American Society of Civil Engineers for the Spring Convention in Atlantic City, New Jersey, on April 28, 1987. The purpose of the session was to focus on this timely topic and provide the practitioners a forum to describe the methods of repairs and inspection of infrastructure. It also provided a forum to promote the exchange of information between the practitioners and the academicians, which may lead to relevant research in the area. This publication contains the proceedings of this session.

The first paper, by Zingher and Stahl, entitled "Williamsburg Bridge Cable Replacement," describes how the pitted cables were replaced while maintaining the bridge in service.

The second paper, by L. R. Nucci, entitled, "Inspection, Repair and Maintenance of Marine Structures," describes the effects of the marine environment on existing marine structures and outlines techniques for inspecting and evaluating the damage to the structures.

The third paper, by C. M. Minervino, entitled, "Rehabilitation of the Pershing Square Viaduct, New York City," describes how the Pershing Square Viaduct was restored under severe traffic.

The fourth paper, by R. Richard Avent, entitled, "Concrete Repairs—Are We Ready to Design Them?," discusses the effectiveness of various methods of repair to concrete structures and points out the need for the engineers to be able to quantify the effectiveness of repairs and thus design the repairs.

The fifth paper, by W. M. K. Roddis, entitled, "Concrete Bridge Deck Condition Assessment: Traditional and Innovative Inspection Technologies," describes how the condition of a bridge deck should be assessed to minimize rehabilitation costs.

The final paper, by M. J. Garlich, entitled, "68th Street Intake Crib—Inspection and Repair," describes details of the inspection and repair of an intake structure.

Each of the papers included in the Proceedings has received two positive peer reviews and has been accepted for publication by the Proceedings Editor. All papers are eligible for discussion in the Journal of Structural Engineering. All papers are eligible for ASCE awards.

The editor wishes to express his gratitude to the authors for their contribution, without whom this publication would not have been possible. The editor also acknowledges the assistance provided by Mr. R. A. Kollmar of Stone & Webster Engineering Corporation, Cherry Hill, New Jersey.

Finally, by no means least, the editor would like to thank Mr. T. Williamson, Member of the Executive Committee of the Structural Division, for his help, and Ms. S. Menaker of the ASCE staff for her cooperation.

Mahendra J. Shah
Editor
February 1987

CONTENTS

WILLIAMSBURG BRIDGE CABLE REPLACEMENT

by Naomi Zingher [1] and Frank L. Stahl [2], F.ASCE

INTRODUCTION

The Williamsburg Bridge is a suspension bridge across the East River in New York City. During a condition inspection in 1979 it was found that the wires of the main cables were heavily pitted due to a general corrosive breakdown of the wires' protective coating. Laboratory investigation determined a considerable loss in tensile strength and an alarming lack of ductility of some 150 wire samples removed from the bridge. As a result of accelerated corrosion testing and computer modeling it was concluded that the cables had a very finite life.

Since replacement of this heavily travelled bridge with a new structure is not considered a viable option because of time, traffic maintenance, financial and environmental considerations, plans are now being prepared for a replacement of the cables while traffic is maintained. The four existing 18-3/4 inch diameter cables will be replaced by two 24-inch diameter cables, each composed of 169 individual prefabricated parallel wire strands. The load transfer from the old cable system to the new cable system will be accomplished by jacking systems and careful application of precalculated load increments to the new and existing suspender ropes and the new and existing tower saddles.

The work involves the development of new design technology but utilizes existing available and proven construction methodologies.

DESCRIPTION OF BRIDGE

The Williamsburg Bridge across the East River in New York City connects the Boroughs of Manhattan and Brooklyn. It was opened to traffic in 1903. The bridge is approximately 7,300 feet long from the Manhattan terminus at the intersection of Delancey and Clifton Streets to the Brooklyn terminus at Broadway and Roebling Street.

The structure consists of a suspended 1600 ft. main span and two side spans of 596 ft. each, a 2,606 ft. approach viaduct on the Manhattan side and a 1,865 ft. approach viaduct on the Brooklyn side. The main span is suspended from two pairs of 18-3/4 inch

---

1) Project Engineer, Ammann & Whitney, 96 Morton Street, New York, New York 10014

2) Chief Engineer - Transportation Division, Ammann & Whitney, 96 Morton Street, New York, New York 10014

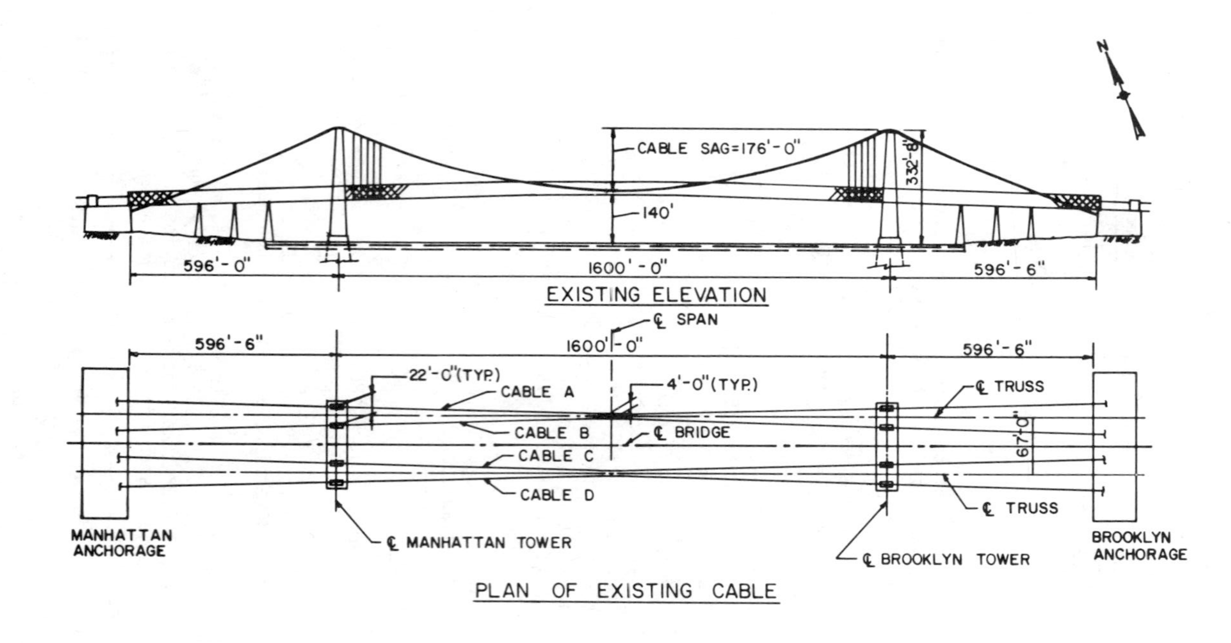
N
CABLE SAG=176'-0"
332'-8"
140'
596'-0"
1600'-0"
596'-6"
EXISTING ELEVATION
℄ SPAN
596'-6"
1600'-0"
596'-6"
22'-0" (TYP.)
CABLE A
4'-0" (TYP.)
℄ TRUSS
CABLE B
℄ BRIDGE
CABLE C
67'-0"
CABLE D
℄ TRUSS
MANHATTAN ANCHORAGE
℄ MANHATTAN TOWER
℄ BROOKLYN TOWER
BROOKLYN ANCHORAGE
PLAN OF EXISTING CABLE

FIG. 1

diameter cables which are supported on two steel towers approximately 335 ft. above water level (Fig. 1). The cables terminate in massive masonry anchorages on both shores. Stability against wind pressure is provided by drawing each pair of cables together at the center, and also by a double system of lateral bracing in the plane of the top and bottom chords of the two trusses.

The two side spans are not suspended from the cables. They form a four-span structure between the main tower and the masonry anchorage, supported by 3 intermediate towers. Therefore, the cables in this area are straight back stays.

Each cable consists of 37 strands of 208 No. 6 gauge (0.192 in) bright (ungalvanized) parallel steel wires. Even though galvanizing was used 20 years earlier on the Brooklyn Bridge to protect the wires, the designers of the Williamsburg Bridge decided against this type of protection apparently because of the slight loss of strength resulting from the galvanizing process. Instead, the individual wires were treated with a proprietary hydro-carbon "slushing oil" compound mixed with 25 percent artificial graphite (1). After all wires had been placed, they were squeezed into a cylindrical shape and held in this form by cast steel clamps which carry the suspender ropes. Between clamps the cables were protected with a wrapping of cotton duck soaked in the hydro-carbon compound and covered with two half-shells of sheet iron. In 1920 this wrapping system was replaced by the now existing helical wrapping of galvanized soft wire.

Originally, the bridge was designed for 2 railway tracks, 4 trolley tracks, 2 roadways and 2 elevated promenades for pedestrians and bicycles. Today, the bridge carries 4 two-lane roadways, 2 rapid transit tracks and one pedestrian promenade (Fig. 2).

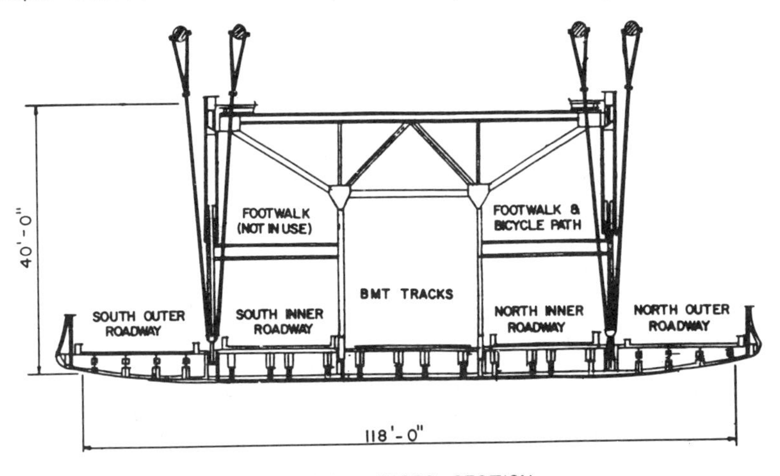

TYPICAL CROSS SECTION

FIG. 2

The most commanding feature of the suspended structure is its two 40-foot deep stiffening trusses. Together with a transverse overhead truss at each panel point, lateral bracing systems at top and bottom chord level, and the closely spaced plate girder floor-beams, they form in effect a deep torsional box of considerable stiffness and wind stability. The main span is suspended from the cables by a unique system of suspender ropes; each suspender is actually a continuous rope connecting its floor panel to both cables of each pair (Fig. 3).

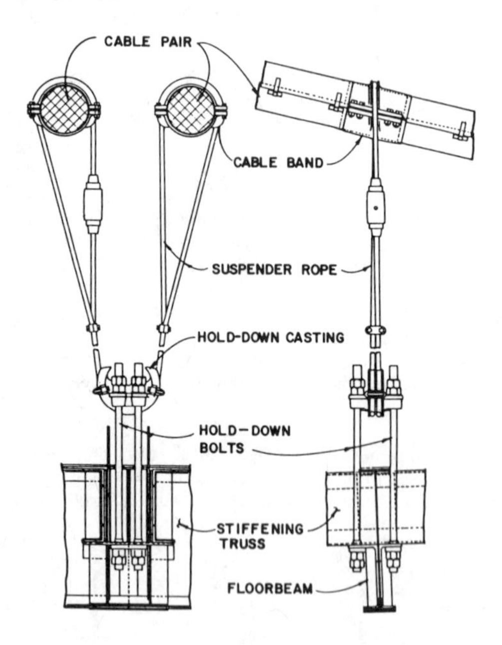

WILLIAMSBURG BRIDGE
ORIGINAL CABLE SYSTEM FIG. 3

The Williamsburg Bridge is of considerable engineering-historical significance. It is the second-oldest of New York's large river crossings. Completed only twenty years after the Brooklyn Bridge, itself an engineering marvel and a milestone in bridge construction, the Williamsburg Bridge bears witness to the rapid industrialization and the great technological advances of the country at the turn of the century. Although the mainspan of the Williamsburg Bridge exceeds that of the Brooklyn Bridge by only five feet, the total length of structure exceeded the Brooklyn Bridge by almost one-quarter mile. The mainspan was at that time second only in length to the Firth of Forth Bridge, and exceeded in capacity any other bridge yet built. In 1903, the Williamsburg Bridge was considered the largest bridge structure in the world. It's total load carrying capacity was about twice that of the Brooklyn Bridge.

In contrast with the Brooklyn Bridge with its limiting terminal stations at both ends, the Williamsburg Bridge incorporated modern ideas of providing a broad, continuous thoroughfare over which trains, vehicles and pedestrians could pass without interruption. The bridge thus became an integral part of the City's street system.

The great progress made in design analysis is apparent in the method of stiffening the suspended floor against deformations. In the Brooklyn Bridge this was accomplished by six shallow trusses assisted by a series of stiffening cables running from the truss panel points to the top of towers - a not entirely satisfactory arrangement as unfortunately proved by the buckling failure of the trusses several years after completion of the Bridge. In the Williamsburg Bridge, stiffness was obtained by the two 40-foot deep continuous lattice trusses, connected at each panel point by transverse plate girder floorbeams and overhead trusses, and by a double system of top and bottom lateral bracing.

Equally great progress had been made in manufacture and construction. The Brooklyn Bridge was built at a time when steelmaking was in its infancy and some of its members were made of wrought iron such as the pin-connected truss members. For the Williamsburg Bridge, high-class steel with riveted connections was used exclusively. The towers were the first built of steel, stone masonry having been used on all major suspension bridges prior to this time. Steel wire, which was used for the first time by Col. Roebling in the construction of the Brooklyn Bridge, was greatly improved in the interim, providing a 25 percent increase in ultimate strength from 160,000 psi to 200,000 psi. Thus, the cables on the Williamsburg Bridge provide 80 percent more carrying capacity than those of the Brooklyn Bridge with only a 45 percent increase in size. Total construction time for the Williamsburg Bridge was seven years as compared to thirteen years for the Brooklyn Bridge.

Another note of historical significance is the fact that the Williamsburg Bridge was one of the last major bridge structures designed prior to the advent of automobile traffic.

INSPECTION AND TESTING

In 1978, the New York State Department of Transportation engaged Ammann & Whitney to perform, for the first time in the life of the Bridge, a complete in-depth inspection of the bridge structure and approaches. This effort entailed an analysis and evaluation of all components of the superstructure and substructure in accordance with present-day standards. The inspection covered the entire steel superstructure, including towers, supporting bents and columns; cables and suspender ropes; the exposed steelwork within the two anchorages; the steel framing supporting the rapid transit system; and all exposed masonry of piers, anchorages, abutments and retaining walls.

All four cables were unwrapped at selected locations and wires were removed for testing. At some locations, the cable was spread apart by wooden wedges to inspect and to remove wires from interior layers. Visual observation of the samples indicated that, generally, the original "slushing oil" protective compound is still tightly adhering to the wire but has, in many places, a rusty appearance. Where the coating was removed with emery cloth, the wire exhibited pitting of varying degree, from a few thousandth of an inch in depth and diameter to one-third of the diameter of the wire. (Fig. 4). Pitting was found on every wire sample removed from the bridge; the degree of pitting did not seem to bear a specific relation to the position of the wire sample in the cable.

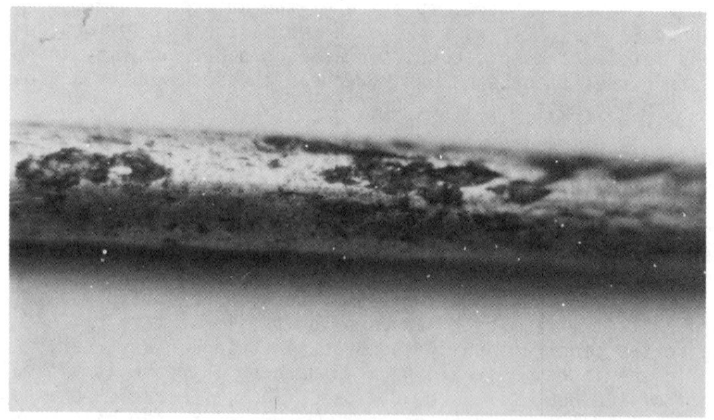

PITTED WIRE SAMPLE **FIG. 4**

Original wire specifications called for a minimum ultimate strength of 200,000 psi and minimum elongation in 8 inches of 5 percent. Final construction reports indicate that the wire as furnished had a minimum ultimate strength of 225,000 psi, with a probable minimum elongation of 3 percent. Physical tests of the wires showed a wide range of breaking strengths, the lowest being some 20 percent below the minimum ultimate strength originally furnished (2). Most elongation values were also below the specified minimum. A statistical probability calculation indicates a probable minimum strength at this time of about 25 percent below the furnished minimum value or about 15 percent below the specified minimum wire strength.

Since the carrying capacity of a cable depends on the uniform distribution of loads between the various wires in the cable and since this capacity to distribute loads depends primarily on uniform ductility of wires, the fact that the initial physical testing indicated very low ductility for many wires caused considerable concern. Consequently, elongation tests were carried out on 300-foot long samples removed from the straight backspan of the bridge in order to determine a ductility more comparable to the actual in-place condition (3). Five of the 18 samples tested had an elongation of less than 1.0 percent. The first one percent of stretch is function of the wire modulus of elasticity and the wire cross-sectional area. This is the elongation in the elastic range. In order to meet specified requirements, the bridge wire must provide elongation in the plastic range. From the results, it was apparent that all wires displayed elongations that were significantly reduced from the specified values.

Fatigue tests showed that, although there has been a loss of fatigue life roughly comparable to the loss of strength, there is sufficient capacity remaining at this time (4).

Based on similar inspection and testing experience on many other suspension bridge cables, the results of the physical tests of the Williamsburg Bridge cable wires not only came as a complete surprise but appeared to put the continued serviceability of the bridge itself into question. Once the extent of wire deterioration became apparent, a further and more sophisticated testing and research program was therefore initiated, with the following main tasks:

- Review the available literature to obtain information regarding the corrosion of steel in the presence of aggressive atmospheric environments.

- Examine wires removed from the cable in detail, using optical and scanning electron microscopy, to assess the principal corrosion mechanism(s) operating on the Williamsburg Bridge cables.

- Perform laboratory experiments on the removed wires under accelerated conditions in simulated environments to determine as closely as possible the current corrosion rates.

- Perform multiple tensile tests on specimens of bridge wire with smooth gage lengths (corrosion removed) to determine a statistically meaningful average breaking force for the wires in the uncorroded condition and to estimate how these breaking forces are distributed through the cable.

- Using the results of the literature review, the laboratory experiments, and the detailed examinations, develop a model capable of making realistic estimates of the present and future load-bearing capacity of the wires in the cable.

This concentrated research effort (5) resulted in several important conclusions and forms the basis for all considerations concerning the future life of the Williamsburg Bridge.

Accelerated Corrosion Experiments. - The wires in the bridge cables experience a constant change from dry to wet environment, not only because of changing seasons but due to constant variations in temperature and humidity which cause condensation inside the cables. Thus, wires will be exposed to, or immersed for periods of time in rain water or condensed water vapor that has become trapped within the cable for extended periods. Accelerated corrosion rate experiments were therefore performed on single and 3-inch diameter bundles of wire samples taken from the bridge. These samples were exposed to wetting and drying cycles in aqueous solutions that simulated acid rain. The composition of these solutions was based on analyses of water collected at the bridge site. The accelerated corrosion tests, both on bundled and single specimens, indicated a current localized and pitting corrosion rate between 0.6 and 2.5 mils/yr of metal penetration. Results of these experiments were incorporated into the cable strength computer model (see below).

Metallurgical Examination and Testing. - To properly ascertain the amount of strength degradation the wires had experienced as a result of corrosion damage, it was important to ascertain the original strength of the wires. A total of 167 tests samples were prepared by carefully grinding the corrosion damage off the wire and thus obtaining material essentially in the same condition as originally installed. These test coupons, with a machined 1-inch gage section, showed a wide variation in strength from wire to wire, ranging from a low of 200 ksi to a high of 249 ksi. However, their grouping into 4 distinct groups of about 210, 220, 230 and 245 ksi suggests that the wires were supplied from several batches and heats produced by somewhat different thermomechanical processing and different composition. The mean ultimate tensile strength of all 167 tests is 227 ksi, close to the value reported at time of cable installation. This confirms that the wires have suffered no degradation other than that caused by pitting and surface corrosion.

Extensive metallurgical investigations of corroded wire samples were performed to (1) determine if the degradation in strength observed in mechanical testing was the result of environmental effects or resulted from metallurgical defects in the wire and (2) determine if in-situ wire failure resulted from simple tensile overstress after sufficient area reduction or was the result of a time-dependent fracture process (i.e. stress corrosion cracking). Observations were made on the fracture surfaces of wire samples tested to failure, fracture surfaces of wires found broken in the cable (in-situ breaks) and the fracture surfaces of the smooth 1-inch gage specimens.

It was concluded that environmental degradation due to pitting and general corrosion was the principal cause of the in-situ failures. Similarly, environmentally induced surface defects and not localized internal metallurgical defects were determined to be the cause of the reduced strength observed in the laboratory physical tests. These surface defects reduce the load-bearing area of the wire and cause the initiation and propagation of the "low ductility" failure process.

All fractures as well as some 18 ft. of randomly selected and longitudinally sectioned wire segments were examined for signs of stress corrosion cracking. The presence of hydrogen-assisted stress corrosion cracking could not be confirmed. However, the possibility that it is occuring somewhere in the cables cannot be ruled out.

Furthermore, even if stress corrosion cracking has not yet occured, it may almost certainly be a factor in the near future if the general corrosion and pitting process currently operative is allowed to continue unhindered.

Conclusions. - Metallographic and fractographic observations tend to support the pitting corrosion strength degradation mechanism. There is no recorded experience of similar pitting corrosion with galvanized bridge wire of a chemistry and strength similar to the wire used on the Williamsburg Bridge. It can therefore be concluded with a reasonable degree of certainty that the pitting corrosion observed on the Williamsburg Bridge results from the interaction of the coating, or more specifically the graphite contained in the coating, with moisture and steel. Since every inch of cable wire is coated with this compound, every inch of cable wire must therefore be suspect. All observations made on the structure support this suspicion.

MODEL FOR CABLE STRENGTH

Since only a fraction (about 4 percent) of the cable wire is readily accessible and it is therefore impossible to physically determine with any great accuracy the distribution, extent, and history of the corrosion damage along the cable and through its cross section, a computer model to predict cable strength was developed. With the help of this model simulation a forecast can be made about the continued safety of the cables.

The strength of the cable at any time is controlled by the number of unbroken, uncorroded wires in the cable as well as the number and average breaking strength of the unbroken but corroded wires contained within the critical length. The number and degree of corroding wires, the number of broken wires, and the average breaking strength of unbroken wires are a function of assumed corrosion initiation and assumed corrosion growth rate. The model thus incorporates the wire strength as originally furnished, the strength results of the recent testing program, results of the accelerated corrosion experiments and reasonable assumptions regarding the number of wires that start to corrode each year.

Based on what are believed to be the most realistic assumptions of a corrosion rate of 1.0 mil/yr and a clamping distance (the distance required for a broken wire to resume its load bearing function due to frictional load transfer with adjacent wires) of 400 ft., the strength degradation model estimates that the present safety factor of the cable (in straight tension) is 2.5 and that it will drop to 1.7 by the year 2005 (Fig. 5). The original design safety factor was 4.6. Consequently, the reduction to the current safety factor, while still acceptable, is significant.

Confidence into the viability of the model was greatly bolstered by the fact that the number of broken wires discovered during unwrapping of the cable to secure the 300 ft. long wire samples showed a very reasonable correlation with the number of broken wires predicted by the model.

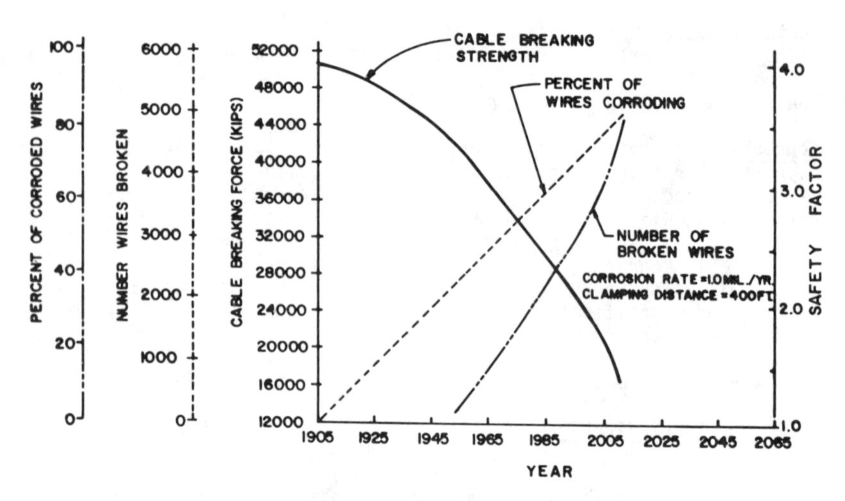

WILLIAMSBURG BRIDGE

BEST ESTIMATE OF CABLE STRENGTH DEGRADATION FIG. 5

SUSPENDER ROPES

Tests performed on two suspender ropes removed from the bridge (2) indicated considerable corrosion, severe loss of ductility and a considerable loss of strength from that originally furnished. Hold-down bolts were also severely corroded.

In order to prevent a catastrophic failure of the bridge due to the accidental breakage of just one suspender rope, every third rope was replaced on an emergency basis in 1983.

REHABILITATION AND REPLACEMENT STUDIES

When it was realized that the cables had a very finite life unless corrosion could effectively be arrested soon, a comprehensive research and development program was initiated to ascertain the viability of cable rehabilitation. The main thrusts of this effort were:

- Basic development studies of new technologies to ascertain existing cross sectional area of the cables, by such means as electro-magnetic, eddy-current, radiographic, acoustic-emission and magnetic flux techniques.

- Basic development studies and tests of corrosion inhibitors and injection techniques to prevent further deterioration. Tests indicated that vapor phase inhibitors were successful in retarding the corrosion rate of wires in relatively small bundles. The ability of inhibitors to permeate all the way through an 18-inch diameter cable remains unknown as does their ability to completely arrest any future corrosion.

- Load testing of the bridge to measure stress levels and deflections. This test consisted of imposing known loads in both symmetric and asymmetric configurations onto the bridge and recording deflection and strain magnitudes. Deflections were measured by a photogrammetric survey and compared with bridge dead weight initial geometries. From the survey data a computer model of the structure was developed to aid in future analysis and design efforts.

- A design effort for a partial stress relief of the main cables in accordance with estimates of existing loss in strength. Alternative schemes included the addition of auxiliary 5-inch diameter suspension cables, the addition of stayed cables, and the possibility of laying new wires directly on top of the existing cables. It was found that utilizing auxiliary suspension cables was viable for up to 20 percent relief. However, the problem of cable deterioration would remain, and no positive means of stopping the corrosion could be relied upon.

Much valuable information has been generated through these studies, but by the very nature of the problem all data remain inconclusive regarding the condition of the main cables. With today's state-of-the-art a reliable determination of the extent of corrosive deterioration of the inner 80 percent of the cables is not possible while they actively remain in place. Based upon the uncertainty of its structural integrity and load carrying capacity there is no doubt that continued reliance on the cable system is not acceptable.

Alternative schemes were therefore investigated to totally eliminate reliance on the corroded cables. Options studied included cable replacement and bridge replacement in its entirety, as follows:

- New Bridge on same location
- New Bridge erected adjacent to and jacked into existing location
- New Bridge in a new location

Various conceptual designs for each new bridge alternative were studied and the engineering feasibility of each alternative, comparisons of community and environmental impacts and associated costs, construction time and construction costs were reviewed. It was concluded that in consideration of the importance of maintaining uninterrupted vehicular and rapid transit traffic on this vital artery, the funds available and the time frame required for the execution of a new bridge project in this location, the construction of a new bridge - while at the same time maintaining the existing

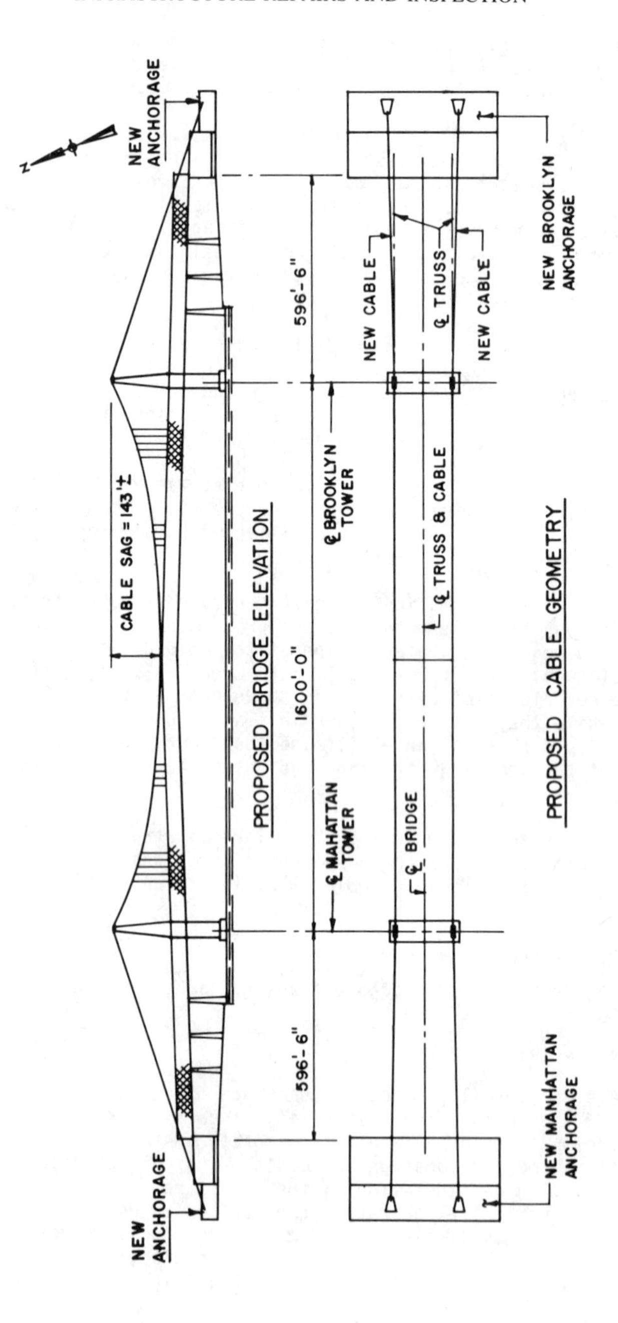
N
NEW ANCHORAGE
CABLE SAG = 143'±
PROPOSED BRIDGE ELEVATION
596'-6"
1600'-0"
596'-6"
℄ BROOKLYN TOWER
℄ MAHATTAN TOWER
NEW CABLE
℄ TRUSS
NEW CABLE
℄ TRUSS & CABLE
℄ BRIDGE
PROPOSED CABLE GEOMETRY
NEW BROOKLYN ANCHORAGE
NEW MANHATTAN ANCHORAGE
NEW ANCHORAGE

FIG. 6

bridge in a safe and serviceable condition - was not a viable option. It is therefore proposed to replace the cables which with rehabilitation of the rest of the bridge will provide the most prudent expenditure of available funds with least disturbance to the travelling public and local communities.

CABLE REPLACEMENT SCHEME

Cables, as the primary support elements on a suspension bridge, have been replaced in some instances in the past. However, in each such case, the bridge was closed to traffic and the suspended structure was at least partially disassembled and later rebuilt. In this case, however, traffic closures are an unacceptable condition and the cable replacement will be designed so that the bridge remains in service; traffic, for the most part, will be fully maintained.

Each existing cable pair will be replaced by a single cable approximately 24 inches in diameter, placed in between the existing pair, directly above the stiffening truss in the center span and at a slight angle to the stiffening truss in the end spans (Fig 6). The sag of the new cables will be approximately 143 feet; the low point will be 8 feet above the top the stiffening truss. The two cables will be constructed of prefabricated parallel wire strands. Each cable will consist of 169 strands of 61 galvanized, 0.196 inch diameter wires with a minimum ultimate tensile strength of 235 ksi.

The new cables will be anchored in new blocks of concrete construction with stone facing behind and bearing against the existing anchorage housings. For optimum anchorage configuration, the cables will angle slightly outward after passing over the tower in order to pass the stiffening truss just before entering the anchorage block. The new anchorage blocks will be founded on piles, depending on soils investigations. Stage construction will be employed to take advantage of the cable pull to reduce foundation pressures.

The new cables will be supported on the towers by new cable saddles placed in between the existing saddles of each cable pair. The load from the cables will be transferred to the tower legs by new girders placed in between and below the existing cable saddle grillage girders. Since some portions of the tower legs are considerably overstressed, under current load conditions, the existing columns will be reinforced by the addition of a third tower column in each north and south face of each tower leg.

To allow for continuous load transfer between the two cable systems with complete and safe load carrying capability for each system at any given time, the new cable system will support the bridge from the top chord, while the existing cable system is supporting the bridge from the bottom chord of the stiffening truss (Fig. 7). The new cables will be attached to the suspended structure by a pair of single leg suspender ropes at each stiffening truss panel point. The new suspender ropes will be galvanized IRWC ropes, pin connected to the cable band and connected with standard

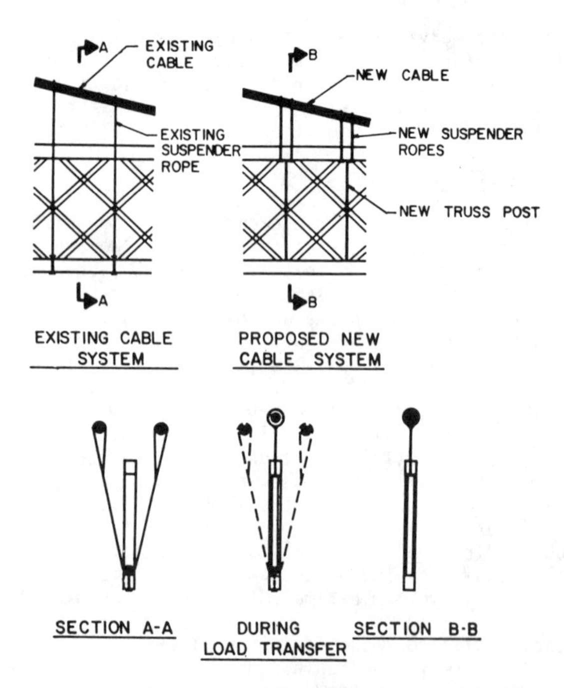

SCHEMATIC CABLE REPLACEMENT CONFIGURATION

FIG. 7

bearing sockets below the top chord of the stiffening truss to a new vertical post. Local reinforcement of stiffening truss diagonals will be required.

Survey has indicated that the present dead load position of the stiffening trusses at the center of the main span is approximately two feet below their original design elevation, due probably to increase in dead loads and creep of suspender ropes. It is planned to restore the stiffening trusses to their original position at completion of the cable system replacement. During load transfer operations, the trusses will deflect upwards about three feet at the center above this final position.

With proper protection, traffic on the structure during construction will be maintained with minimum interruptions. However, restrictions to rapid transit usage during cable erection and load transfer may be necessary.

LOAD TRANSFER PROCEDURE

The load transfer from the old to the new cables must be accomplished by a carefully controlled jacking procedure (Fig. 8). This procedure has to satisfy the following requirements:

- Gradual and uniform tensioning of the new cables and detensioning of the existing cables.
- No overstressing of the cables, suspenders, trusses and towers during the transfer.
- No cable slippage through the saddles.

The load transfer operations involve the installation of temporary jacking frames at the tower tops for each of the existing and new saddles, and at alternate panel points for the new suspender ropes. These latter frames will be attached to the suspender ropes with Lucker cable grips and will load the ropes by jacking against the stiffening truss top chord, similar to the system used very successfully in the suspender rope replacement for the Golden Gate Bridge (6).

The new saddle connections to the tower tops will be designed to permit movement only during the load transfer. The existing saddles, supported on rollers, were designed to allow movement during normal bridge operation. Recent investigations concluded that the existing saddles are "frozen", i.e. the unbalance in the horizontal cable tensions caused by live load and temperature is not able to overcome the friction force and to cause relative displacements between the saddles and the tower tops. For the load transfer the force required to overcome this friction between the saddles and tower tops will be supplied by jacks.

During the load transfer, old and new saddles will be jacked in opposite directions, simultaneously, to avoid large movements of the tower tops and unacceptable additional stresses in the towers. The new suspender ropes will be tensioned simultaneously by jacking against the top chords of the stiffening trusses.

In order to keep erection stresses in the suspended structure and towers to a minimum, cable erection and load transfer must proceed simultaneously for both cables. The load transfer will be performed in two stages, each consisting of several intermediate steps. Basically, Stage I consists of stressing the cable and the new suspender ropes sufficiently so that they can be connected to the stiffening truss. In Stage II, the existing suspenders will be unloaded so that they can be disconnected from the truss.

Stage I. - The initial (no-load) position of the new cables at midspan will be 20 feet higher than their final position; and the new tower saddles will be positioned 2.75 ft. shoreward of their final position. The load will be transferred to the new cable by reducing the distance between the new cable and the truss, while the distance between the existing cable and the truss remains constant.

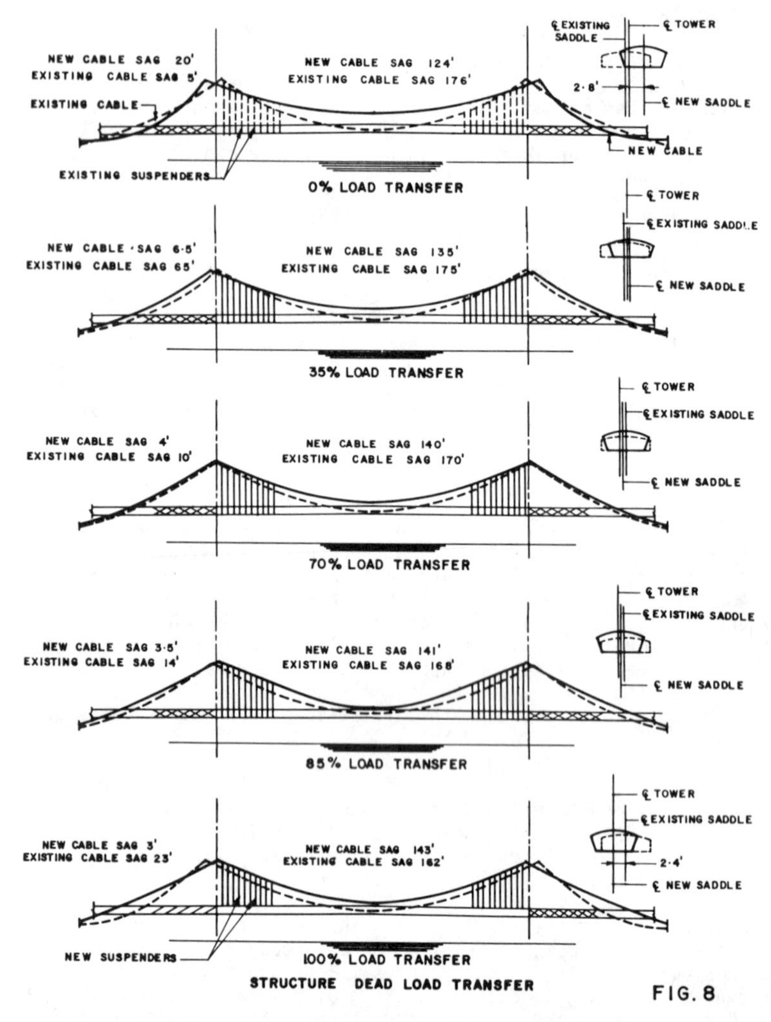

STRUCTURE DEAD LOAD TRANSFER

FIG. 8

Fifteen jacking steps, of 2 to 7 percent load transfer each, will be required to connect at first every other new suspender to the truss at alternate panel points. Each step consists of two consecutive operations:

> The new saddles will be jacked riverward and the existing saddles will be jacked shoreward to precomputed positions. This disturbs the equilibrium of the horizontal cable tensions for each system.

- The new suspenders at alternate panel points will be jacked to precomputed tensions. This equalizes the horizontal cable tensions for each system.

One jacking step of less than one percent load transfer will then be required to connect the remaining alternate suspenders to the truss.

- The new suspenders at the alternate panel points will be jacked and connected to the truss.
- The new and existing saddles will be jacked to equalize the horizontal cable tensions for each system.

At the completion of Stage I about 70 percent of the structure dead load will have been transferred from the old cables to the new cables. At that time, the trusses will be deflected upwards about three feet at the center above their original design position.

Stage II. - Opposite to the procedure for Stage I, during this stage the load remaining in the old cable is transferred to the new cable by gradually increasing the distance from the existing cable to the truss while the distance between the new cable and the truss remains constant.

One jacking step of less than one percent load transfer will be required to remove two-thirds of the existing suspenders, maintaining the one-third which was replaced recently. After the removal of these unloaded suspenders the new and existing saddles will be jacked to restore equilibrium in each cable system.

The remaining old suspenders will be detensioned in a similar manner in a carefully phased operation until all bridge load has been transferred from the old to the new cables.

Strand-by-strand removal of the old cables in reverse order of their original construction will complete the operation.

Live Load and Temperature Effects. - The load transfer will be performed at night when the flow of traffic on the bridge is very low. After each step of load transfer the saddles will be fixed to the tower tops by the jacking frames. No relative displacement will occur between the new and existing saddles and the tower tops under live load action.

Due to the load stiffening effect on suspension cables the stiffness of both cable systems relative to each other and to the truss is constantly changing throughout the load transfer process. Therefore, the percentage of the live load supported by each cable system varies with each step of load transfer.

The unbalance in horizontal cable tensions caused by live load and temperature could cause cable slippage at certain stages of the load transfer. To prevent this from happening, restrictions to rapid transit usage might be necessary.

Conclusions. -

- The complexity of the load transfer procedure is a direct result of the non-linear interaction between the two cable systems and the stiffening truss.
- Each step of the load transfer requires a trial and error analysis of the structure behavior. The analysis is performed using a finite element non-linear computer program on sophisticated mathematical models of the structure.
- Large tower saddle movements (about 3.0 ft.) caused by large changes in backstay sag will require full control through the load transfer.
- The amount of dead load transferred will have to be small enough to prevent the unbalanced cable tension to cause slippage of the cables through the saddles.
- The live load and temperature effect has to be considered for each step of the load transfer.
- The procedure developed takes into account the interaction between the existing and new elements and insures a gradual transfer of the load without interrupting the traffic on the bridge.

SUMMARY

1. Inspection, tests and research have determined that the ongoing pitting corrosion in the wires of the Williamsburg Bridge cables has greatly reduced their useful life. Pitting corrosion was traced to an interaction between the wire and its protective coating and must therefore be assumed to exist throughout the entire length of all four cables. No way has been found to arrest this corrosion and the cables must therefore be assumed to have a very finite life.

2. Detailed studies have established the feasibility of replacing the present cable system with a new cable system while maintaining the bridge in service.

3. Cable replacement can be performed using presently available materials and construction procedures of proven history.

4. Cable replacement, when combined with a general bridge rehabilitation and coupled with a continuing maintenance program, is a viable means of restoring full integrity to the Williamsburg Bridge for the next 100 years. Reconstruction can be achieved by tried and proven construction methodologies. This option has the shortest construction time, least negative community impacts and can be accomplished within available funding constraints. Although

occasional lane closures will be required, multi-directional traffic flow can be maintained through all phases of construction. Also, the historic countenance of New York City's second oldest bridge will be maintained as an important historic engineering monument.

ACKNOWLEDGEMENT

The Williamsburg Bridge is owned and operated by the City of New York. All work reported herein has been performed under a contract with the New York State Department of Transportation , and supervised by both the City and State Transportation Departments.

APPENDIX - REFERENCES

1. Wilhelm Hildenbrand, project engineer for John A. Roebling's Sons Company, ENGINEERING NEWS, November 13, 1902.

2. W. T. Gore and H. W. Peterson, "Williamsburg Bridge," Final Report, Contract No. UZ-00065, Submitted to Ammann & Whitney Consulting Engineers by U.S. Steel Corporation, Trenton, New Jersey (September 17, 1980).

3. W. A. Lucht, "Evaluation of Elongation Ductility of Main Cable Wires Taken from Williamsburg Bridge," Final Report, Contract No. UZ-01590, Submitted to Ammann & Whitney, Consulting Engineers by U.S. Steel Corporation, Trenton, NJ (October 5, 1982).

4. W. A. Lucht, "Fatigue and Elongation Tests on Wire from Williamsburg Bridge", Final Report, Submitted to Ammann & Whitney, Consulting Engineers by U.S. Steel Corporation, Trenton, New Jersey (May 10, 1982).

5. L. E. Eiselstein and R. D. Caligiuri "Corrosion Damage to the Williamsburg Bridge Main Suspension Cables", Final Report, Submitted to Ammann & Whitney, Consulting Engineers by SRI International, Menlo Park, California (May 1985).

6. Golden Gate Bridge, Highway and Transportation District, Golden Gate Bridge Contract XXXIX "Suspender Replacement and Floorbeam Repairs" and Contract XL "Suspender Replacement - Phase 3", prepared by Ammann & Whitney, Consulting Engineers.

INSPECTION REPAIR & MAINTENANCE OF MARINE STRUCTURES

By Louis R. Nucci, P.E., A Member ASCE*

ABSTRACT

Marine structures are subject to both extremes in loadings and deteriorating forces throughout their life. These exact a toll on the structures reducing the load capacity and decreasing the serviceability of the structure. During their life, the structures have possibly been subjected to normal loadings as well as overloading from wind, wave, and impact forces; corrosion; scour and erosion; biological deterioration; and rot. This deterioration generally occurs below the top structure of the marine facility and, therefore, out of sight from the occasional observer and may even be out of sight from all except those in specialized equipment. On that basis, many times maintenance and repairs are not performed until serious structural distress has been noted.

This paper outlines the effects of the marine environment on existing marine structures, techniques for inspecting and evaluating the damage to the structure, and methods for evaluation of the structural capacity of the structure.

INTRODUCTION

Marine structures have always been of vital importance in the operation and commerce of all nations. They have been constructed of stone, timber, concrete, and steel and are designed and located in areas of extremes of loadings and severe environmental constraints. Marine structures encompass a variety of forms and functions including piers, wharves, bulkheads, breakwaters, bridge foundations, and fender systems as illustrated on Figure 1. All of these must be designed to carry their service loads as well as loads from wind and wave forces. They are subject to a greater extent as compared to land based structures from corrosion, abrasion and erosion, biological deterioration, and rot. With time, they slowly degrade the structure and reduce its ability to carry the service load.

Although the normal life expectancy of a marine structure is on the order of 25 years, marine structures 50, 75, and over 100 years old are still in service providing berthing for vessels and movement of commodities. Because of the activities which occur at these structures, it becomes important to keep these facilities in operation and maintaining their capacities. Since loss of the water interface portion of the overall facility may restrict use of the entire facility, it becomes important to maintain the structure at a serviceable level.

*Consulting Engineer. 1 Middle Road, Merrimacport, MA 01860

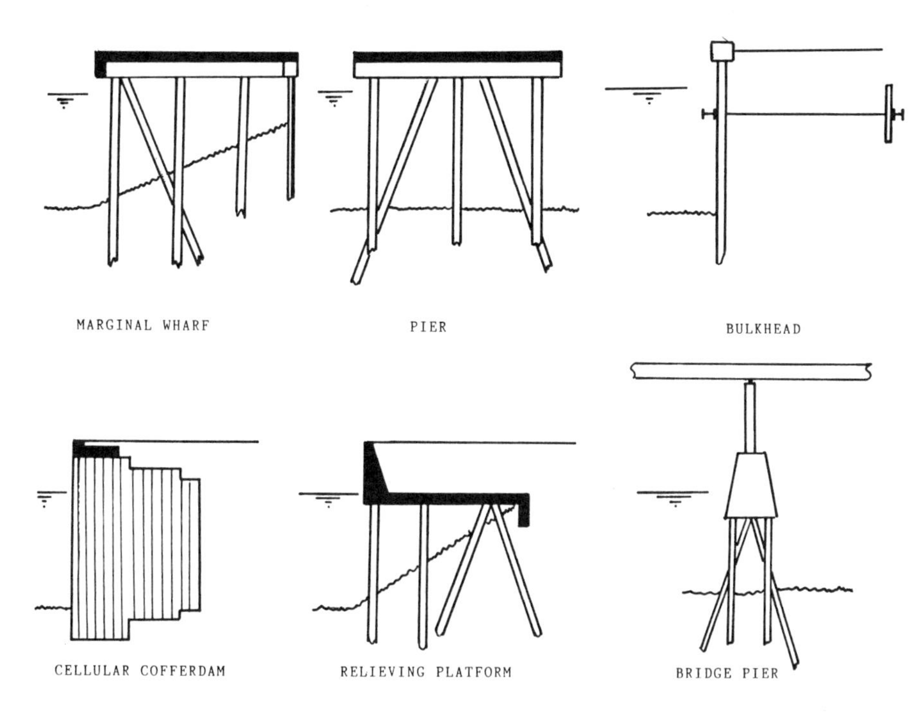

Figure 1 Marine Structures

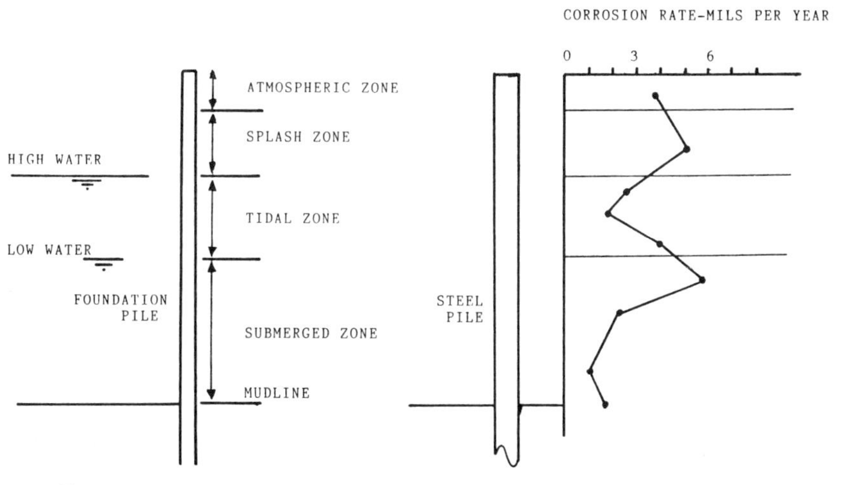

Figure 2 Immersion Zones

Figure 3 Corrosion Attack

The damage and deterioration with the waterfront structures, be it a bridge foundation or pier, generally proceeds either in or under the water surface and underneath the structure. For this reason, the casual observer cannot fully inspect the structure. An indepth approach to the inspection and evaluation of the structure is needed. This includes documentation of the original construction, inspection, analysis and capacity rating, and the implementation of a maintenance and repair program. Agencies such as the Departments of Transportation, Port Authorities, and U.S. Navy and U.S. Army Corps of Engineers have now initiated programs for the inspection, repair, and maintenance of the structures.

With the increased costs of work around and under the water, it is more cost-effective to provide periodic maintenance to the structure rather than extensive repairs. In association with this maintenance, it is also possible to upgrade the facility or improve the service capacity of the structure.

The deterioration of the structures involves degradation processes such as corrosion of steel structures, chemical deterioration of concrete structures, marine border attack, and rot of timber structures. Damage is also caused from overloading structures from impact loads from vessels, as well as wind and wave forces exceeding design considerations. Inspection procedures to evaluate the level of deterioration vary depending on the location and type of inspection. It is done generally by divers, either commercial/construction divers or engineer/divers. Advantages and limitations of each are discussed later. Other inspection procedures may include small tethered submersibles or larger personnel-carrying submersible vessels.

Assessment and repair programs must consider the changes the structure has undergone since construction. The usual change in the structural elements is a reduction in the member cross section resulting in an associated reduction in load capacity. Load transfer mechanisms may bridge deteriorated elements maintaining capacity of the structure, however, at a reduced overall safety factor. Use of two and three dimensional finite element models have been successfully used illustrating this effect and can assist in the determination of critical structure elements. A rating program can be established which determines the critical members and effects of repairs on the overall structural capacity. Repairs can then be accomplished on the "weak-link" of the structure, and maintenance programs established which economically upgrade the structure with each repair procedures.

MARINE DETERIORATION

The evaluation of marine structure deterioration must consider the natural processes specific to the marine environment and the affects these processes have on the structures. The continuing loadings from winds, waves, currents, ice, and ship impact are forces which the structure must sustain while deterioration from corrosion, erosion, scour, biological deterioration, rot, and fatigue occurs.

The deterioration which may occur is dependent on the structure type, construction material, and the location within the marine environment. The degree of deterioration depends on climate; seawater properties; and seasonal variation, tidal ranges, and conditions with respect to the relative immersion of the structure. Figure 2 illustrates relative immersion zones.

The atmospheric zone tends to contain small amounts of salt which increases the rates and amounts of corrosion as compared to steel further inland. In timber structures, freshwater may collect on the members initiating rot of the timber elements. The splash zone is an area above the high water level which continued salt spray. This wetting and drying leads to high levels of steel corrosion and concrete spalling. The tidal zone is the range between mean high and mean low water. This zone is periodically immersed, generally on a daily basis. Corrosion of steel and spalling of concrete occurs in this zone. Additionally, scour and abrasion of structural elements from ice and debris occur within this zone. Below the low tide line is an area where continual immersion occurs. In general, this is an area of light attack to steel and concrete structures, however, it is the area of greatest deterioration of timber structures from marine border attack.

The natural processes creating corrosion, concrete deterioration, scour, marine border attack, and rot are briefly discussed in the following paragraphs:

Corrosion

Marine corrosion is the deterioration of metal by its reaction with the environment. Comments herein will pertain to corrosion of steel structures. The corrosion process is an electro-chemical process which degrades the steel by the production of a metal oxide, generally called rust. Being an electro-chemical process, it requires a current carrying medium between different parts of a corrosion cell. In the marine environment, this medium, the electrolyte, is seawater. The remainder of the corrosion cell consists of elements of the steel structure. A more detailed description of the corrosion process is presented by Uhlig (1).

Many types of corrosion occur in the marine environment. The most common types are:

- Galvanic Corrosion
- Stray Current Corrosion
- Differential Environment
- Erosion Corrosion
- Biological Corrosion

Galvanic corrosion occurs from the connection of two dissimilar metals in an electrolyte (seawater). Current flows through the electrolyte from the more reactive metal (anode) to the less reactive metal (cathode). This results in the corrosion of the anodic area while protecting the cathodic area. Corrosion attack is illustrated on Figure 3, with examples of galvanic corrosion being:

- New steel is anodic to old steel
- Weld metal is anodic to basic steel
- Clean cut surfaces (i.e., pipe threads) is anodic to uncut surfaces
- Highly stressed areas are anodic to unstressed areas

Stray current corrosion occurs where stray direct current passes through the structure to the electrolyte. This current can arise from subway systems, poorly grounded welding generators, and crane systems. The current drives and accelerates the corrosion process leading to an increased corrosion rate.

Differential environment corrosion occurs when localized changes in the environment causes differing electrical potential across structural elements. Generally, differing oxygen levels in the seawater creates the corrosion cell, but it can also be caused by differing salinity. On steel piles, differential oxygen levels just below the low tide level causes high levels of corrosion. Increased corrosion can also occur in cracks, crevices, bolt heads, and corners due to differing environments.

Erosion-corrosion is the result of scouring action of sediments exposing clean metal, thereby keeping the corrosion active. It also removes the corrosion by-product which may have assisted in protecting the structure. Erosion-corrosion is a mechanical-electrical form of galvanic corrosion.

Biological corrosion is caused by marine organism which accelerate corrosion by affecting the environment. The organism may create differing oxygen environment lending to differential corrosion; they may create corrosive products through decomposition; or they may remove protection corrosion by-product film, leading to galvanic corrosion of the steel.

Concrete Deterioration

Concrete and concrete structures show good resistance in the marine environment. Damage includes frost and/or scaling attack, abrasion and erosion, reinforcement corrosion, and chemical attack. In general, deterioration of concrete structures is a combination of the various modes of damage. Typically, levels of deterioration are illustrated on Figure 4.

Frost and scaling attack occurs when water within the concrete surface freezes, causing damage to the inner concrete structure. Concrete decks which are heavily salted during the winter months to remove snow and ice and structures located in the splash and upper tide zones are most susceptible to frost and scaling attack. The presence of any chloride within the concrete impairs its resistance to frost activity.

Abrasion and erosion deterioration is a mechanical form of damage from ice, debris, and suspended sediments. Pilings and other concrete structures near the water can be damaged in areas subject to ice movements. Additionally, currents may create erosion and abrasion of concrete structures near the bottom by movement of granular sediments.

Reinforcement corrosion is the most common mechanism in the deterioration of concrete structures. Corrosion of the reinforcement is primarily due to chloride penetration of the concrete. This results in expansion of the reinforcing steel surface ultimately spalling the concrete surface. Non-watertight concrete, insufficient cover, and cracks in the concrete promote corrosion of the reinforcement. Chemical attack, such as high sulfate concentration in the seawater leads to deterioration of concrete by damaging and softening the concrete and allowing loss of section through abrasion, erosion, and spalling of the concrete.

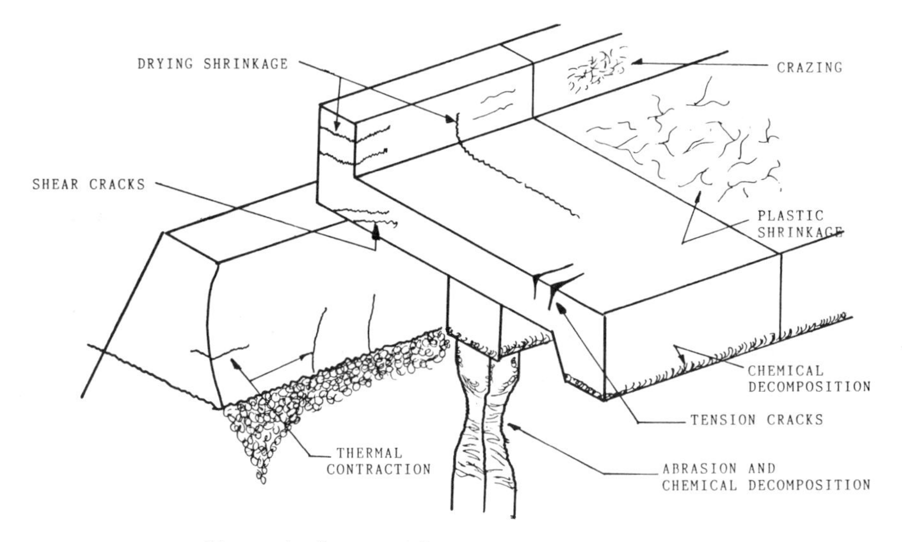

Figure 4 Concrete Deterioration

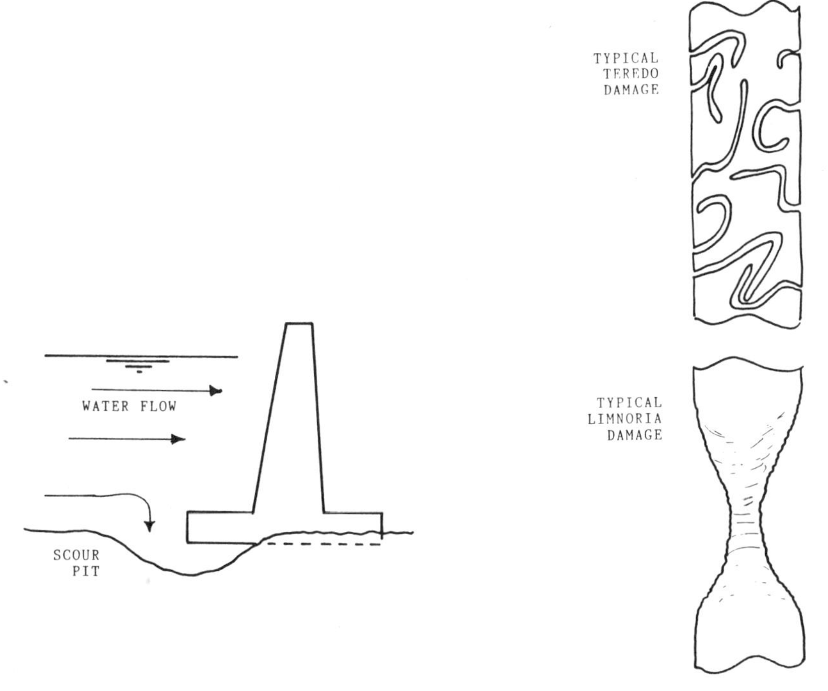

Figure 5 Scour

Figure 6 Marine Borer Attack

Scour

Scour is the removal of the sediments from the base of the structure. The greatest impact from scour is bridge foundations in streams and rivers. This is illustrated on Figure 5. Damage depends on the sediment character in the streambed, volume and speed of water transport, and the shape and size of the structure. Hydraulic design of bridge and marine structures is well advanced, and usually structures are designed properly for scour protection. However, stream, oceanographic, and structural conditions may change during the life of the structure creating increases in scour.

Design of breakwaters and solid fill structures may encounter scour depending on wave and current regimes during storm conditions. Major storms can quickly move vast quantities of bottom material resulting in deeper water at the structure than original design considerations. Differences in design water depths based on uniform literal drift and extreme storm conditions may result in undermining of the structure. Undermining of bulkheads and seawalls are also commonplace when the effects of wave reflection on the solid structures are neglected.

Marine Borer Attack

Marine wood borers are a highly destructive group of crustaceans that cause damage worldwide. The more common borers are the mollusk Teredo (shipworm) and Pholads, these both being burrowing borers, and the crustacean Limnoria (Gribbles) which damage timber by cutting furrows into the surface of the member. Borers and types of damage they inflict are illustrated on Figure 6.

Old harbors contain large numbers of timber pilings as well as other wood structures. Because of their age, the piles were either installed unprotected or the creosote protection has been lost due to natural leaching processes. Although these piles have been at risk to borer penetration for years, increases in borer attack has not occurred until recent. Prior to the 1970
had been held in-check by harbor pollution.
an improved water quality in many harbors, providing an environment for the borer activity to flourish.

The Teredo is probably the most destructive of the marine borers. They enter the timber at the wood surface and bore into the timber as it grows. External evidence of the Teredo is a tiny entrance holes which are difficult to observe. As the Teredo bores, the interlacing tunnels of many Teredo eventually hollow the pile, destroying the load capacity of the pile.

The Limnoria are surface gouging borers, cutting furrows in the softer portion of the timber. This creates a lacy network of tunnels and furrows at the surface seldom more than 1/4" deep. This weakened surface is eroded by wave and current action exposing new wood to the borer. As this continues, a necking down of the timber occurs, eventually destroying the timber.

The Limnoria are generally found in the lower energy zones of the marine environment. Larger wave impact tends to pull the borer off the wood surface. If, however, the Limnoria can burrow behind boltheads, into check and cuts, or behind bracing, they can hollow out the pile without little external evidence.

Rot

Timber structures are susceptible to attack by various types of fungi, generally known as rot. Two major types of rot are prevalent on marine timber structures, white rot and brown rot. White rot is visible as a white bleached appearance of the wood surface. Brown rot (dryrot) is visible as the wood surface is eaten away, leaving a brown residue.

To exist, the fungi needs air, water, and food. Areas which allow for ponding of freshwater are highly susceptible to rot. This includes checks and cracks in tops of stringers and pile caps. As rot progresses, it provides a natural basin for the water to collect. By preventing water from collecting and providing adequate ventilation, the incidence of rot should be greatly reduced. Wood that is always dry or always wet seldom rots.

INSPECTION PROCEDURES

An important part of the inspection and rehabilitation of waterfront structures is the physical assessment of the existing conditions of the structure. The inspection must be carefully planned and implemented to obtain the greatest amount of data from the inspection. The initial phase of any inspection program is the collection of all available documentation on the structure. This can include project plans and specifications, as-built drawings, permit and license applications, public accounts of construction, and any maintenance and repair records concerning the structure. Much of this documentation is generally unavailable for older structures with increasing available documentation with younger structures. Agencies, such as Port Authorities, major industrial facilities, Defense Department Agencies, and the U.S. Coast Guard, have better documentation of their facilities. As the level of documentation decreases, it becomes more necessary to obtain more physical data during the field inspection.

Geotechnical investigations are also important in the inspection and evaluation of existing marine structures. This is particularly important if there is any evidence of settlement or movement of the structure. If suspected, a geotechnical investigation should be performed to determine if the subsoil has undergone changes which affects the structure.

Upon completion of the data collection and geotechnical engineering analyses, an inspection program is developed to evaluate the existing conditions of the structure, the level of deterioration, as well as the mechanisms which led to the existing conditions. A properly planned condition survey is an important key to any rehabilitation project. This leads to a survey where all necessary information is obtained. Since surveys are both costly and time consuming, planning provides the client with an accurate and cost efficient documentation of the condition of his structure.

Inspections are generally conducted in two parts. The first part is an above-water survey, consisting of all visible members which can be accessed without specialized equipment. The second portion is an underwater survey, conducted by divers and/or submersible video systems. The two surveys should be coordinated to insure that the elevation where one survey ends is the same as where the other begins. Often it has been noted where the above water survey inspects the structure above high water while the underwater survey inspects the structure below low water. Because of this lack of coordination, conditions within the tidal zone were not noted.

The above water survey should include a layout of the existing structure; detailing pile locations; member size, location, and frequency; connection details; splice locations; and other pertinent information. Surveys should include lengths and elevations, as well as physical locations of all other elements. Documentation should include physical measurements, sketches of details, and photographs. Photographic documentation should include both general conditions as well as unusual conditions. The full range of the structure condition should be noted. This is important since evaluation of the structure should be based on average load capacities where the general conditions were noted with reduced capacities in areas of more severe deterioration.

The underwater survey part should be completed, if possible, in two surveys. The first survey is a quick inspection of the structures below water. This survey should provide a sufficient amount of data on the structure condition to perform a pseudo-statistical analysis of the conditions. This allows a preliminary determination of the general structural capacity. It also provides direction for the final survey to concentrate on areas of greatest deterioration.

The final survey is a complete survey of the underwater conditions. Inspections cover the full range of the structure from tide zone to the mudline. A proposed sequence of inspection areas is shown on Figure 7. The tide zone is covered with multiple inspection locations, with documentation concentrated immediately below the low tide zone. Uniform spacing of documentation in the immersed zone is performed to provide efficient data gathering. Increased inspection near the mudline is used to evaluate increased potential for abrasion type damage.

There are four primary methods for performing an underwater survey. They include:

- o Construction/Commercial Divers
- o Engineer/Diver
- o Remote Sensing, Tethered Video Systems
- o Diving Bell or Submersible

The most common method employed is the use of commercial/construction divers. They are available in nearly all waterfront locations. To be properly performed, the underwater survey must be clearly defined for the divers. Locations and types of measurements should be delineated by the engineer directing the survey. It is important that the engineer constantly review data gathering by the divers. Level of documentation of existing conditions should also be defined to insure that the level and extent of deterioration is properly delineated.

Engineer/divers are becoming more widely used in the underwater surveys of marine structures. Many agencies now require the inspection be conducted by engineer divers, most notably the U.S. Navy. Engineer divers should be experienced in the design methods for marine structures and in the analysis of

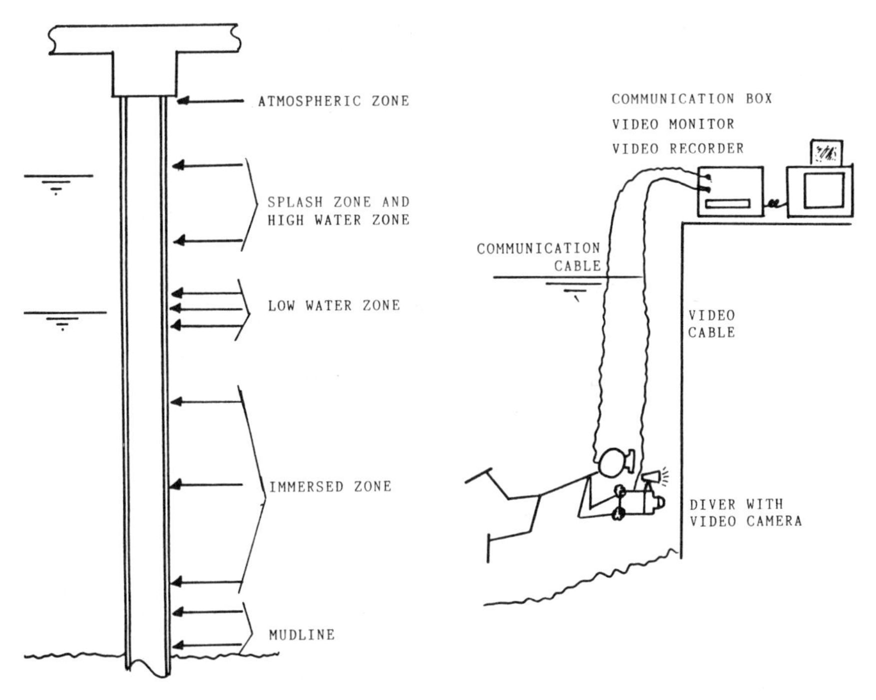

Figure 7 Inspection Ranges

Figure 8 Underwater Video

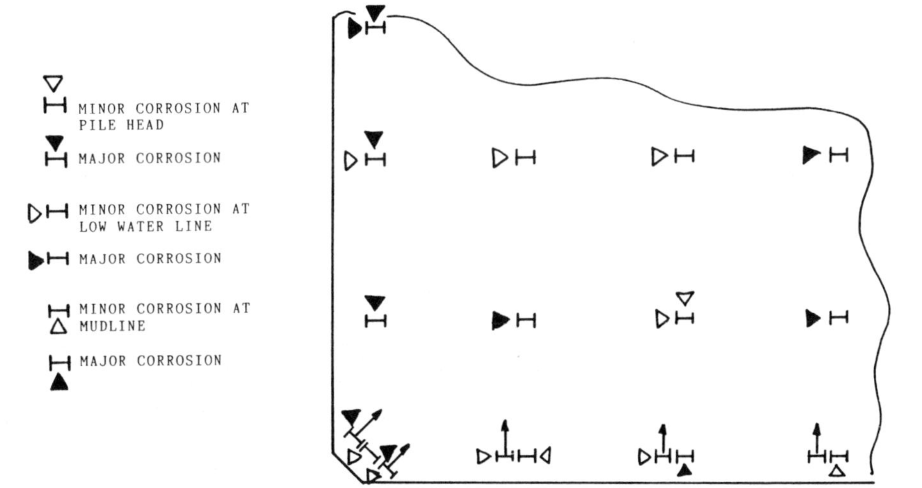

Figure 9 Typical Inspection Plan

existing structures. Use of engineer divers provide a "trained eye" which evaluates levels of deterioration and effects on the structure. The effects of deterioration, load transfer, structure movement, repairs, and modifications to the structure require evaluation; and the best person to provide this is personnel who participated in the survey.

Remote sensing-tethered video systems are becoming used more frequently for visual inspections of marine structures. In areas of high risk, deep water, or long penetration dives, the tethered video systems become very economical. Where water clarity permits, these systems allow continual visual inspection without concern for diver safety or decompression. The tethered system consists of an underwater video camera incorporated into a remote control submersible. A system is illustrated on Figure 8.

Submersibles and diving bells are used primarily for deep dive inspections for structures such as oil platforms. They have limited use in nearshore marine structures and will not be discussed in this paper.

Inspection equipment varies depending on the structure type and construction material. The equipment ranges from visual documentation including photographs and video equipment to non-destructive measuring devices including calipers, tapes, ultrasonic thickness meters, to destructive testing including increment borers, coupon cutting, pile slicing, and removal of members.

A listing of many of the various inspection devices are described on Table I.

Final results in the inspection of a waterfront structure is the preparation of a factual condition report. The report should include a listing of all surveys made and indicate areas where inspection was not possible. Dates, times, and weather conditions should be carefully noted for the inspections. For example, light to moderate marine borer activity noted in early Spring may become heavy borer activity by late Summer with attendant increase in timber damage. Surveys much greater than 6-12 months old loose their usefulness as conditions can change quickly as deterioration advances. Graphical representation of the conditions noted should be presented. Photographs should be keyed to the structure by descriptive or graphical locations. As much as possible, the graphics should diagramatically illustrate degree of deterioration. A typical pile plan with damage notations is illustrated on Figure 9.

EVALUATION

The purpose of evaluating the existing conditions is to determine the service load capacity and to design repairs necessary to maintain and/or upgrade the facility. The analyses must consider the live load capacity of the structure, including the ability to sustain both vertical and lateral loading. A rating type approach is proposed to develop capacity in terms of individual structural member capacity and overall facility capacity.

The evaluation of the structure is conducted in two phases. The first phase is an engineering analysis to determine service load capacity. This includes review and analysis of the original construction and the effects of deterioration on the structure. The second phase is the development of an urgency index, which will depict the urgency of corrective action and relate how local conditions are affecting rates of deterioration. This index can also be used in maintenance programs for restoration of the facility.

To accomplish this requires use the available documentation for the structure and the results of the inspection program. For the basic structural elements, original design capacity and existing reduced capacities are developed.

Load capacity determinations are made for the foundations, substructure, and superstructure. As with most structures, the actual design capacity is the lowest capacity strength of the components of the structure. For example, the design capacity for vertical loads of a pier might be 600 psf. With design considerations and construction practices, the deck could have been designed for a capacity of 650 psf, the pile caps of 600 psf and the pile layout of 750 psf. Similarly, design load capacities for lateral and seismic forces can be determined for each element of the structure.

The effects of deterioration of the structure are now considered. For each structural element, the reduced capacity is developed. If deterioration is relatively uniform, the reduced capacity of the members is developed as an average capacity loss. This reduces the computations necessary to develop the remaining capacity. Allowable member loads would be developed for the piles, substructure, and superstructure with the remaining service load being the minimum of the allowable member loads.

If the deterioration is not uniform throughout the structure, an alternative approach is to subdivide the structure in sections based on the general layout of the structure. This could be by pile bents, deck panels, stringer spacing, or some other repeatable pattern based on the structural system. Again, allowable loads for foundations, substructure, and superstructure are computed based on the condition of the elements within the section. A mosaic of the allowable structure capacity would be developed with the weakest structural element identified. This presents a graphical description of the facility capacity and illustrates the element which governs the capacity. This approach is very similar to bridge rating programs in which allowable load capacities are governed by the "weak link" structural element in the bridge structure. Typical results of this approach are illustrated on Figure 9.

For example, a timber pier as shown on Figure 11, consists of piles, pile caps, stringers, bracing, decking, fender systems, and miscellaneous structures. Design capacity is based on the capacity of each of these elements working as a unit. The design vertical service load capacity is the maximum live load that could be applied without overstressing any of the structural members. Original design capacity is 400 psf vertical loading. Based on an inspection program, average section loss for the elements is as illustrated in Table 2.

INSPECTION METHODS	DESCRIPTIONS
VISUAL	DIVER OBSERVATIONS PHOTOGRAPHIC DOCUMENTATION UNDERWATER VIDEO RECORDING
TACTILE	DIVER'S HANDS CALIPERS TAPE MEASURES SURVEY EQUIPTMENT
NON-DESTRUCTIVE TESTING	ULTRASONIC THICKNESS MEASURING DEVICES MAGNETIC PARTICAL CAT SCANNING
DESTRUCTIVE TESTING	INCREMENT BORERS CORING CUTTING OF SLICES PULLING OR REMOVAL TEST-PIT EXCAVATIONS

UNDERWATER INSPECTION DEVICES
TABLE 1

STRUCTURAL MEMBER	INITIAL CAPACITY (psf)	AVERAGE LOSS OF CROSS SECTION	CAPACITY (psf)
PILES	550	40%	300
PILE CAPS	500	30%	350
STRINGERS	400	25%	250
DECKING	700	25%	550
FASTENERS	N/A	50%	N/A

TYPICAL TIMBER PIER STRUCTURE
TABLE 2

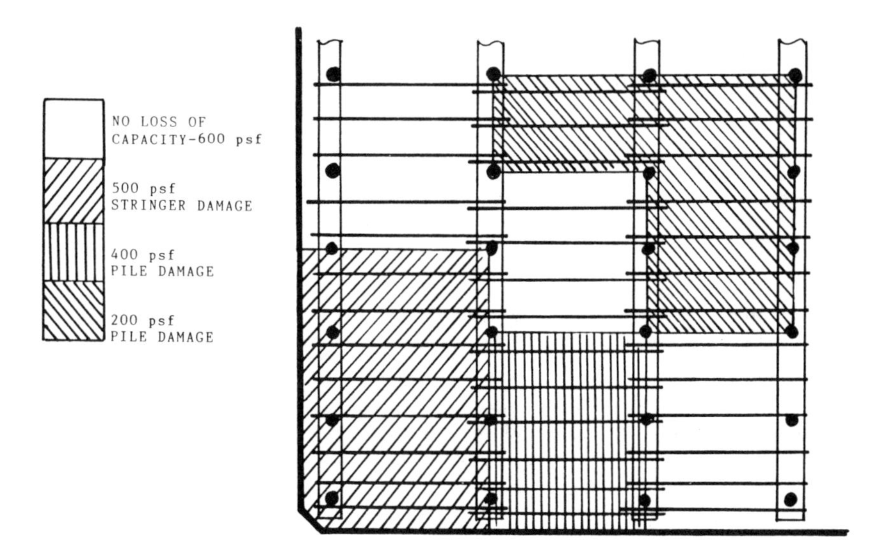

Figure 10 Inspection Plan Mosaic

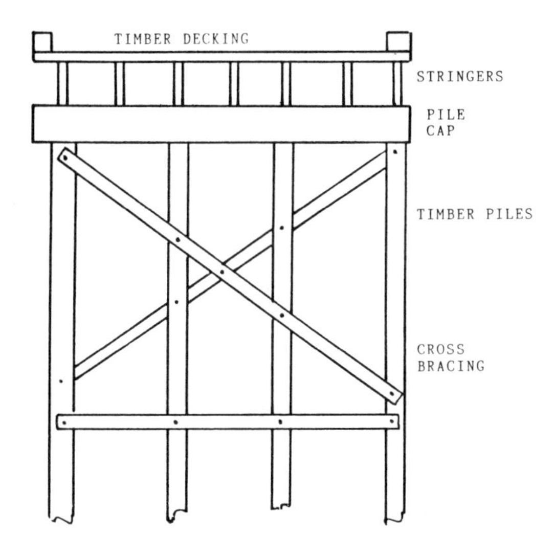

Figure 11 Typical Timber Pier

Loss of section of the pile affects both the compression capacity and buckling loads. Loss of section of the pile caps, stringers, and deck affects bending capacity and shear capacity. In addition, a review of cap and stringer horizontal shear capacity should be performed. It has been found that this generally governs the remaining strength of the structure. Often it appears that this analysis was not performed for older structures, and the capacity is governed by this. Based on the evaluation, the allowable load capacity for this pier is 250 psf. This approach also indicates that for this pier, the stringers govern the allowable loads on the structures, although the greatest amount of deterioration is occurring at the piles. Repairs to upgrade the facility must first concentrate on the weak element to provide any increase in capacity for the structure.

Using the same techniques, lateral load capacity of the structure is also evaluated. Effects of deterioration of the structure on vessel impact and berthing and vessel mooring can be developed and the weak link identified.

The next phase in the evaluation of the structure is the development of an urgency index. The urgency index is used to evaluate the existing condition of the structure and the potential for continued deterioration. This system is effective for use in maintenance programs, since it utilizes regular assessment of the structure.

The intent of this index is to maintain the facility at a current load capacity. As the deterioration reduces the capacity of the members, the rating decreases. The urgency index is shown on Table 3.

A rating of 9 indicates the member is in a "like new" condition. No disconcernable deterioration has taken place. An 8 is essentially the same except that deterioration processes have begun. This may include corrosion staining of steel or noticeable cracks in concrete.

A rating of 6 to 7 indicates that deterioration is occurring with no significant reduction in member capacity or overall structure capacity. For example, steel piles may be designed for sacrificial loss of cross section for corrosion to occur. This is expected and does not affect the capacity.

Ratings of 4 to 5 indicate that the structure has undergone deterioration to the extent that the allowable capacity is below the design capacity. It may also indicate significant section loss to elements which do not at this time affect the desired capacity. At level 5, corrective action should be completed within the current construction season. At level 4, the work becomes a priority item.

Ratings of 2 to 3 indicate capacities of below that desired. At this point, it is necessary to reduce the allowable loading of the facility. The structure would thus become "posted." Reduced capacities of the facility mandate repairs be implemented as soon as practical.

An index of 1 or 0 requires the closing of the facility. This is generally indicative of an emergency repair to the facility.

Maintenance Urgency Index	Conditions Noted Based on Maintenance Rehabilitation Level	Maintenance Urgency Definitions Based on percieved Need for Rehabilitation
9	New Condition	No Work Required
8	Good Condition	Work Required Only for Cosmetic Repairs.Items Should Be Tracked For Future Maintenance.
6 to 7	Maintenance Required	Items Should Be Considered For Maintenance work During Next Work Cycle. Repair Type is Probably Class I Repairs.
4 to 5	Rehabilitation Required	Allowable Load Capacity is Below Design Capacity. Rehabilitation Work Should Be Performed Within Next Year or Sooner. Class II Repair
2 to 3	Major Repair Required	Load Capacity of Structure is Severely Restricted. Repairs to Be Accomplished As Soon As Possible. Class III Repair.
0 to 1	Facility Closed	Facility is Unsafe For Intended Operations.Major Reconstruction Necessary.

URGENCY INDEX RATING SYSTEM *
Table 3

* Modified from NCHRP Report #251 (2)

MODIFICATION	DESCRIPTION
+2	No Threat For 5± Years, Conditions Have Stabilized, and Condition is Cosmetic in Nature.
+1	No Threat For 2 to 3 Years, Conditions Are Worsening Slowly, and Has Little Structural Effect.
0	No Immediate Threat, Conditions are Deteriorating at Normally Expected Rate, and Deficiency Does Not Lead to Reduction of Structure Capacity.
-1	Threat likely within 1 Year. Conditions Are Worsening and Are Causing Reduction in Structure Capacity.
-2	Threat is imminent. Conditions Are Worsening Rapidly and Severe Loss of Structure Capacity is occuring.

Note: Modifers cannot increase rating above '8'.

MODIFICATION TO URGENCY INDEX SYSTEM *
Table 4

The rating can also be modified based on the engineer's/inspector's assessment of the deterioration. Modification to the existing urgency index can be on the order of two steps in either direction. The modifications are shown on Table 3. For a deficiency which is stabilized or does not pose a threat for 5 years, a maximum upgrade of 2 is proposed. Upgrades are proposed when the deterioration condition is cosmetic or part of a redundant structural system. The use of an upgrade is also for a condition not expected to change within the next 3 to 5 years.

Reductions to the indices may be warranted when it is anticipated that the deterioration can worsen quickly or may have a significant effect on the structural capacity. Conditions which tend to accelerate with time, such that it is expected that prior to the next inspection period the index would have decreased, warrant a reduction modification.

The use of the index system provides a method for tracking the condition of the structure over several maintenance periods. With the detailed evaluations of structural capacities, it assists in determining those areas which require repair and maintenance. It also assists in the allocation of maintenance budgets by directing the funds to correct the elements which have the greatest affect on the structure capacity and use rather than simply on the elements showing greatest deterioration.

REPAIR TECHNIQUES

Many techniques and systems are available for rehabilitation of marine structures. The design and development of the repair programs are dependent on the particular structural element, construction material, and the level of deterioration. As the number of repair techniques are extensive, only representative repair techniques will be discussed herein. The elements most often involved in rehabilitation are:

- Foundation Piles
- Pile Caps and Deck Structure
- Bulkheading

Repair techniques for these will be illustrated in the following sections. Repair techniques can be broadly classified into three generic categories. These categories are defined by the level of deterioration and the magnitude of loss of structural load carrying capacity. The classes are:

- Class I - Deterioration Barrier Systems
- Class II - Structural Reinforcing
- Class III - Structural Replacement

The first repair class acts as a barrier against further deterioration. They primarily encase the element from the environment forces which cause deterioration. These repairs are generally completed as a maintenance item. They extend the structure's life, arrest deterioration, and limit subsequent loss of service load capacity. These repairs are considered to be non-load supporting.

The second repair class involves strengthening of the existing structural elements. These repairs are performed on members which, due to deterioration, cannot support their original design capacity. A rehabilitation program may involve placement of stiffeners, sisters, additional members, encasements, or other techniques which supplement the existing member capacity. Upon completion, full service load capacity would be restored.

The third repair class is for those members which no longer support load. A repair program of partial or complete replacement of members would be performed. The transfer mechanism of load through the structure would be interrupted while the deteriorated member was excised and a new member placed. Full service load capacity of the structure could be realized.

Foundation Piles

Foundation piles are generally the elements most subject to deterioration. Piles extend from the atmospheric zone to the submerged zone and below the mudline. Construction material includes steel, concrete, and timber. Figure 12 illustrates typical repairs with various construction materials for the three classes of repairs.

The most common Class I repairs to foundation piles are the initial treatments applied prior to their installation. For steel piles, these include coal tar and high solids epoxies, fusion bond coatings, and sheathing. For timber piles, the use of creosote and chemical preservatives are also considered Class I repairs. For concrete piles, the use of Type V cement for sulfate resistance would also be a Class I repair.

For installed foundation piles which have been subject to deterioration, many Class I type repair systems are now in use. For steel piles, various coating systems, including high build epoxies and fast drying painting systems, can be applied in the atmospheric, tidal, and submerged zone. These systems can be applied after the steel has been subject to corrosion. The most important factor in these coating applications is the surface preparation. To provide satisfactory protection, coating systems require good cleaning to obtain complete bonding. For steel pipe piles, a PVC wrap also provides satisfactory corrosion protection. Another system is the placement of an epoxy encasement on the piles. A fiberglass jacket is placed around the pile and a thin (1/4" -1/2")(6 mm-12 mm) epoxy poured in the annulus.

Operation of a cathodic protection system is also considered a Class I repair. A properly designed, installed, and maintained system provide protection to all members below the waterline. The selection of an impressed current system or a sacrificial system depends on structure layout, cost, ability to maintain, and life expectancy.

Class I type repairs for timber piles are now accomplished by use of a variety of PVC wraps. Zippertube and Pile-gard are two examples. The PVC is tightly wrapped around the timber pile. This prevents oxygenated water from reaching the timber. Thus borer activity cannot exist on the pile surface. Placement of concrete in fabric forms along the pile or

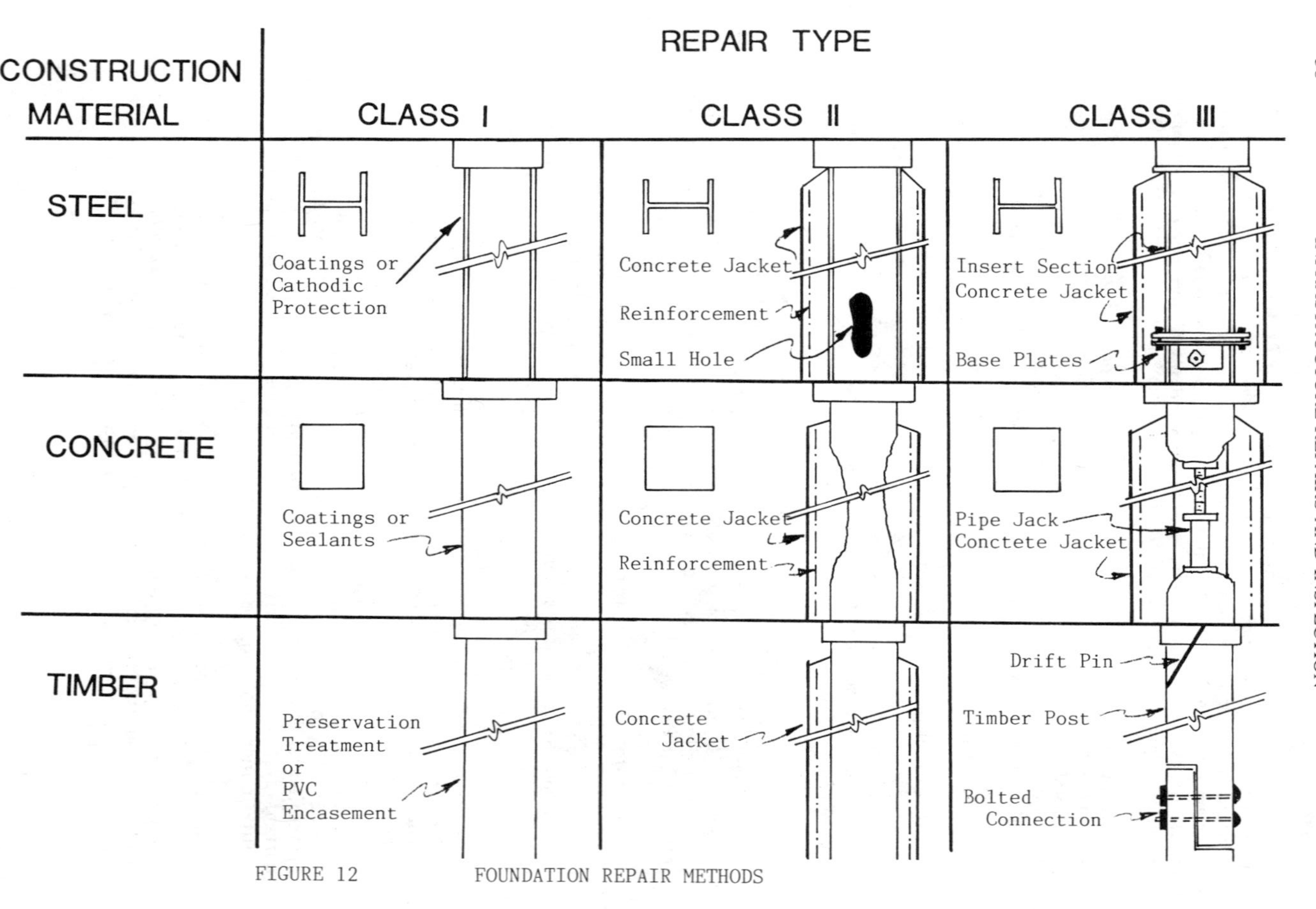

FIGURE 12 FOUNDATION REPAIR METHODS

pneumatically applied in the tidal and atmospheric zone also will arrest deterioration. Loss of pile capacity due to the imposed concrete encasement weight should be evaluated.

Class I type repairs for concrete piles include placement of PVC wraps, epoxy coatings, and rubberized asphalt tape. These repairs arrest chloride penetration into the concrete or reducing the effects of scour or erosion of the pile at the mudline or the tidal zone.

Class II type repairs are arrest any further deterioration and re-establish the load capacity of the pile. The most common system for steel piles is the placement of a reinforced concrete jacket in the zones of deterioration activity. As necessary, additional steel including plates, angles, and/or channel sections could be attached to the piles to provide needed steel section and stiffness. The combined section could be evaluated taking advantage of the increased stiffness of the jacketed pile.

For timber piles and concrete piles, reinforced concrete jackets are also used in re-establishing service load capacity of the piles. For all jackets, pile surface cleaning is very important in the performance of the pile-jacket system.

Selection of the jacket form is important for the success of the project. Fabric forms are easily installed and lowest cost of the forms, however, in areas of high currents or placement on battered or intersecting piles, they tend to loose their shape. Rigid fiberglass forms are more difficult to install but can be easily mass produced. Special rigid forms can be provided to allow for intersecting piles or encasing the pile cap.

Class III type repairs for foundation piles re-establish the load capacity of the pile. For those piles which no longer provide support to the structure, it is possible to drive piles adjacent to the affected pile and establish contact with the pile cap. Unless the piles are short, this is the most costly repair scheme. Posting of piles is the more common approach to the repairs. The deteriorated section is removed and contact re-established between the pile and pile cap. This is accomplished by placement of a pile insert or post. The post should include a mechanism for restressing the pile. This insures good contact between the pile, post, and pile cap. It also allows for dead load to be reintroduced to the pile. It may also be used to re-level the deck, if necessary.

For steel piles, posts can be constructed of steel sections equivalent to the original pile. Bonnets and bearing plates facilitate installation. For timber piles, new timber pieces can be fitted on the pile and sistered to the pile and cap. Steel pipe jacks with concrete jacket encasements have also been successfully used for posting timber piles. For concrete piles, steel pipe jacks are also used with concrete jacket encasement to repair fully deteriorated piles.

Pile Caps and Deck Structure

Pile caps and deck structures are generally located above the tidal zone in the splash or atmospheric zone. Exceptions include relieving platform structures which the superstructure is located within the tidal zone and bridge piers, where the bottom of the structure is located below the tidal zone. Figure 13 details typical repair techniques for pile caps and deck structures.

Class I repairs to pile cap and deck structures are similar to those for foundation piles. Initial treatments prior to construction include coating systems for steel structures, treatments for timber structures, and sealants for concrete structures. Careful attention must also be placed on design of details to prevent areas more likely to foster deterioration.

For existing steel structures, Class I repairs consist of cleaning and recoating of the various members. Coal tar and high solids epoxies, urethanes, and countless other coatings are available. The most critical aspects in placing coatings is surface preparation and contractor experience. Without proper cleaning, the coating will not adhere. This limits the anticipated protection. The contractor should also have experience in placing the coating system. If that is not possible, a technical representative from the coating manufacturer should be onsite full time until the contractor can properly place the coating system.

For existing timber structures, Class I repairs generally include placing additional preservation treatment as necessary. These are sprayed or painted onto timber members in areas where the preservation treatment is leaching out.

For concrete structures, sealants and overlays are used to protect the concrete surface. Various epoxy and acrylic overlay coatings are available which can be applied to the pile caps and decks. When using these systems, compatibility with the concrete structure is important. Bond strength, flexural strength, and thermal expansion coefficients must be balanced so the combined system will not destroy itself during changing seasons.

Bituminous concrete overlays have been used to provide protection, however, in areas of high salt water concentrations, such as in washdown areas on fishpiers, extensive concrete deterioration can occur below the overlay without damaging the overlay or being observed.

Class II repairs for steel structures include structural reinforcing providing additional members or encasements. Proper attention to connection details is required to insure reinforcing acts compositely with the existing members. Placement of new members adjacent to existing members or sistering reduces the service load imposed on the members. Concrete encasements provide additional stiffness and load capacity to the deteriorated members. Again, the effects of the encasement weight should be carefully reviewed prior to their placement.

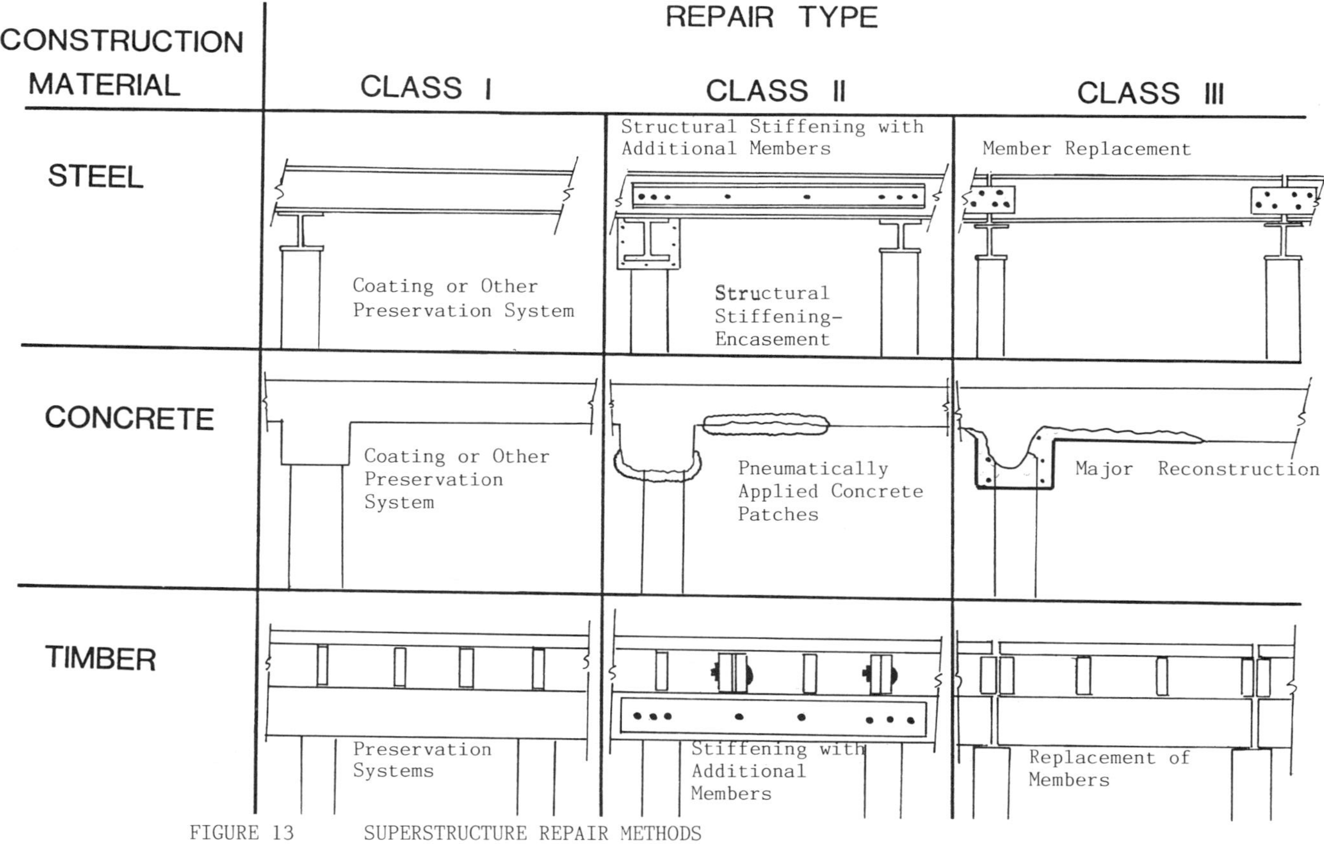

FIGURE 13 SUPERSTRUCTURE REPAIR METHODS

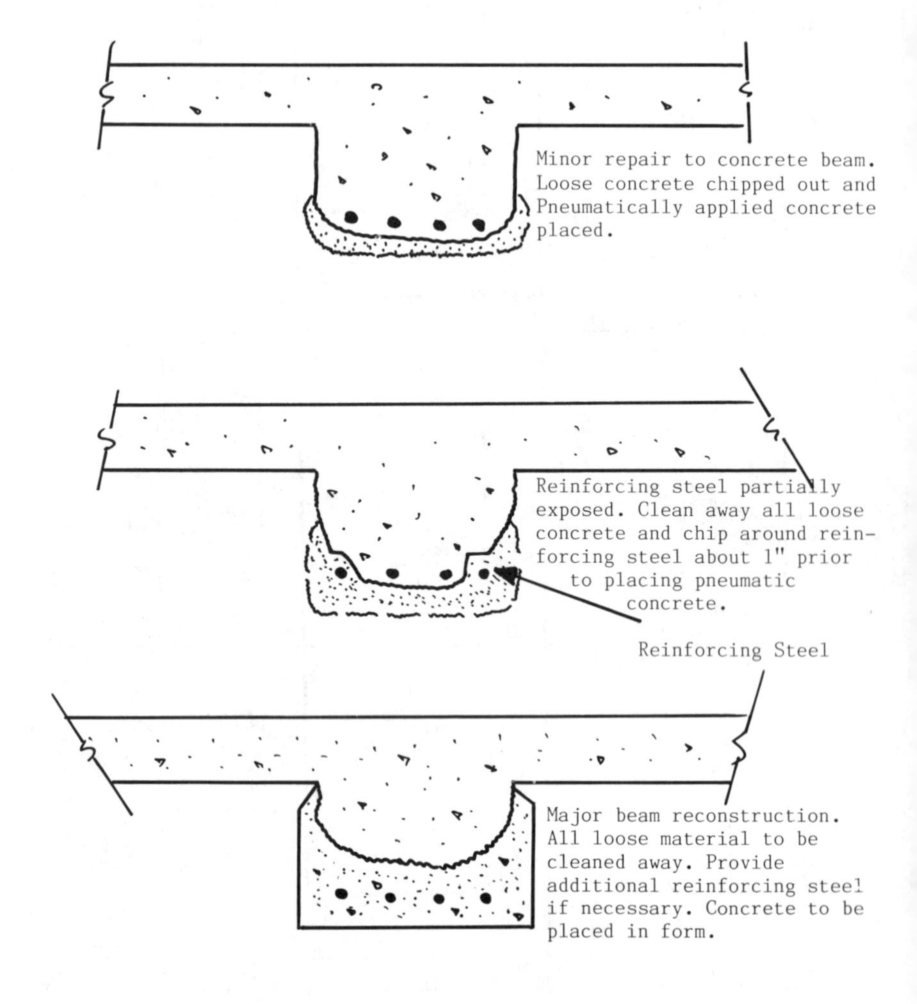

FIGURE 14 REPAIR DETAILS FOR CONCRETE STRUCTURES

Class II repairs for timber structures generally consist of placing additional members or sisters to assist in load support and transfer. New members are placed adjacent to existing members, connected and supported, and then shimmed to support the service load.

Class II repairs for concrete structures include the replacement of deteriorated concrete and protection of the reinforcing steel. For cracked members, epoxy injection and sealing is common. The cracks are injection with a high strength epoxy grout which bonds and seals the crack. Prior to specifying this or any other repair technique, the reasons why the damage has occurred should be determined. The mechanism causing the damage must be understood to insure that the repair will satisfactorily perform.

For larger spalls and breaks, various repair techniques are available. To re-establish concrete cover, pneumatically applied concrete can be used to build up the concrete section. Details are shown on Figure 14. Deteriorated concrete is chipped out, reinforcing steel cleaned, and pneumatically applied concrete placed. Use of bonding agents appears to improve the performance of the finished repair.

Deck spalls on the surface can be repaired by removal of the deteriorated concrete, cleaning of reinforcement, and placement of concrete, modified concrete, or epoxy of acrylic compounds. It is important that all loose and/or deteriorated concrete is removed and the reinforcement and concrete surface cleaned. Bonding agents as recommended by the manufacturer should be used.

Class III repairs for steel concrete and timber structures are most easily accomplished by removal and replacement of the deteriorated members. Review of construction staging and costs should be made. It may be found that more extensive repairs are less costly than localized repairs. For example, in a timber pier stringer replacement program, it may prove advantageous to also replace the decking since the contractor can work from the pier surface rather than below where tides, close conditions, and poor access limit the contractor's production.

For all Class III repairs, proper design of connection details and load transfer is important. The limits of reconstruction must allow load transfer between elements and constructability of the repair. If possible, the repairs should provide uniformity in contractor operations and sufficient space for the contractor to provide temporary support systems.

Bulkheads

Bulkheads, like foundation piles, can be situated in all of the marine environment deterioration zones. Deterioration of steel sheet piles occur at connections and at the corners in the sheet section. Additionally, due to abrasion and erosion, deterioration is more pronounced at the outer face. Concrete sheet piling deteriorates similar to concrete foundation piles. Soil also tends to leak through interlocks. Timber sheet piling deteriorates in a similar fashion to timber piles. Repair details are shown on Figure 15.

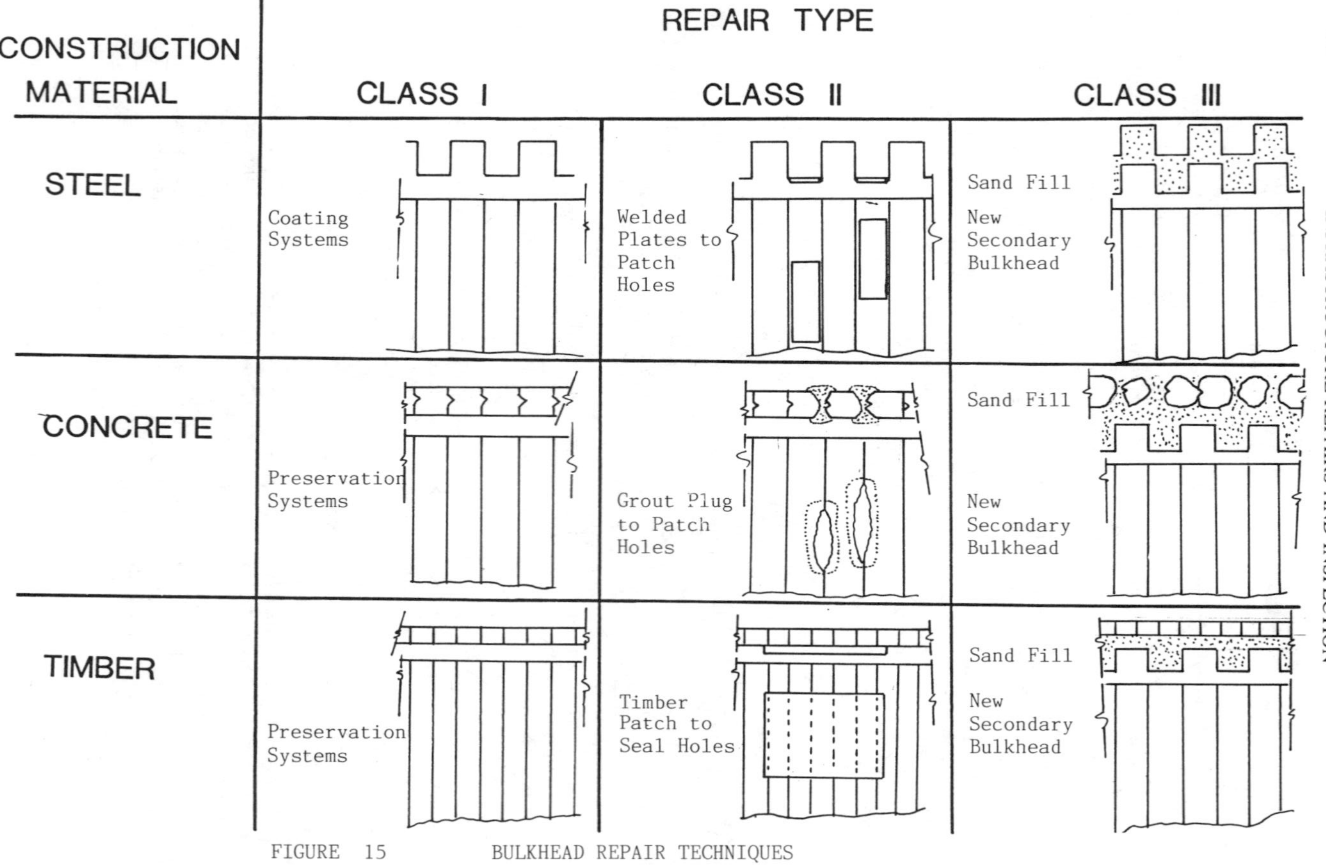

FIGURE 15 BULKHEAD REPAIR TECHNIQUES

Class I repairs for bulkheads are similar to the various foundation pile repaires. These include coatings, sealants, and preservation systems. Maintenance repairs during the life of the structure include recoating of steel piles and cathodic protective systems.

Class II repairs are generally performed when the bulkhead wall becomes perforated and soil backfill is being lost. For steel sheet piles, repairs include sealing of the holes by welding steel plates or placing concrete plugs. After sealing the holes, the backfill can be grouted to further seal the holes.

Grouting of backfill is more often used with concrete and timber bulkheads to seal between adjacent sheets. The implementation of a grouting system must seal the backfill and any voids.

Class III repairs to bulkheads are required when service load capacity behind the bulkhead is reduced. This may include placement of a secondary bulkhead in front of the wall with connection to existing anchorage systems. Fill between the existing sheeting and new sheeting can be either granular or concrete fill.

MAINTENANCE

Maintenance is the process where work is performed to keep a structure at a desired service level. To be effective, maintenance should be conducted on a periodic basis. The inspection and assessment program detailed in a previous section lends itself to a defined maintenance program. Elements are inspected, rated, and compared to previous inspections. Repair work would be defined based on the result of the inspection. Elements of a maintenance program include:

- Inspections and Assessments
- Documentation
- Maintenance Work and Minor Repairs
- Major Repairs

Implementation of the maintenance program must be done on a periodic basis. The frequency depends on age, general condition, construction material, and environment. Annual visual assessments should be undertaken to assess the overall condition. This might entail a walking tour of topside structures and a small boat trip below the structure. Representative photographs should be taken to document existing conditions. Detailed physical measurements and underwater surveys should also be conducted on a periodic basis. Using the urgency index system described before, facilities with ratings of 6 or above can have detailed inspections scheduled at a two to three-year interval. Facilities with ratings below 6 should increase inspections to annual surveys.

Documentation of inspections and assessments is necessary to track changing conditions. A manual providing inspection and assessment guidelines, reporting forms, and long-term tracking assists in the maintenance program. Properly assembled, the manual provides detailed directions for plant personnel to perform the inspection procedures. The inspection forms, when completed, become a part of the manual as reference for the next inspection period.

Maintenance work and minor repairs is the key to any maintenance program. The owner needs to budget maintenance functions into the annual costs of the facility. Periodic and preventative maintenance generally is cost effective over the long-term as compared to major repairs. This annual maintenance cost is only a small fraction of the cost of the facility or the commodities which cross the facility. Any disruption of service due to major repairs is much more costly.

Major repairs which can be predicted by the maintenance program can be programmed into the operation of the facility. Damage of this nature might include continued deterioration of piles or pavements nearing their useful life. This type of repair can be tracked and budgeted as the conditions warrant. Repairs such as collision damage cannot be predicted.

By tracking conditions, owners are not surprised by severe deterioration conditions occurring which require emergency work. Using a maintenance system to track conditions, perform minor repairs, and budget for major repairs will provide a useable marine facility which is economical and safe.

REFERENCES:

1. Uhlig,H., editor. Corrosion Handbook, John Wiley and Sons, Inc. New York, 1948.

2. National Cooperative Highway Research Program, Assessment of Deficiencies and Preservation of Bridge Substructures Below the Waterline, Report No. 251, Transportation Research Board, October, 1982.

Rehabilitation of a Landmark Structure
The Pershing Square Viaduct
Park Avenue, New York City

Charles M. Minervino, P.E.*

The Pershing Square Viaduct carries Park Avenue from 40th to 46th Street in midtown Manhattan, New York City. Constructed in 1917 - 1919, the structure is an elevated drive that crosses 41st and 42nd Streets, makes a circuit around the Grand Central Terminal and the Pan Am Building, crosses 45th Street and descends through the Helmsley Building to ground level at 46th Street. The terminal, viaduct, and many surrounding buildings in the Grand Central zone are representative of Beaux-Arts planning in New York City. The Pershing Square Viaduct with its steel arch spans, granite facing, stone balustrades, ornate cast iron lamps, ornamental cast iron panels and foliate railings, has been designated a Landmark structure by the New York City Landmarks Preservation Commission.

In the early 1900's, the original planning for the Grand Central Terminal complex included a viaduct access structure. The station was constructed first and opened in 1913, with access provided from ground level via 42nd Street, Vanderbilt Avenue and Depew Place. Design of the viaduct was prepared by the architectural firm of Warren and Wetmore with construction work done between 1917 and 1919. At this time the section from 40th to 42nd Street was completed with the easterly terminus ramping down to Depew Place for access to the newly constructed Commodore Hotel and the U.S. Post Office, and the westerly viaduct tying to Vanderbilt Avenue. In 1927, modification to the viaduct included raising the roadway at Depew Place and Vanderbilt Avenue, and constructing the northern portion of the viaduct to its present configuration.

The viaduct is an outstanding example of Beaux-Arts Civil Engineering. The design exhibits French influence, having low rise steel arches with relatively massive granite piers. The approach ramp is granite faced and flanked by a stone balustrade. The arch spans are lined with ornamental foliate railings. Lighting is furnished by cast iron lamp posts set along the viaduct. In recognition of its special history and architecture, the Pershing Square Viaduct was designated a Landmark Structure by the Landmarks Preservation Commission of New York City.

* Principal, Lichtenstein Engineering Associates, P.C., 21 West 38th Street, 14th Floor, New York, New York 10018

Today, the viaduct serves as a major traffic link between upper and lower Manhattan carrying high volumes of cars and taxis. The roadway provides access to a number of important commercial buildings and hotels, as well as Grand Central Station. The areas under the viaduct, fronting along 42nd Street and Vanderbilt Avenue, are enclosed and house a number of retail shops and commercial offices. The roadway deck serves as the roof of the shops with only a suspended ceiling inbetween. The condition of the viaduct had deteriorated, and leakage through the deck disrupted the retail businesses. Our firm was engaged by the New York City Department of Transportation to perform in-depth inspection, rating, evaluation and design of structure rehabilitation, and to provide inspection of construction.

In general, the existing superstructure throughout was composed of riveted steel girder and floorbeam construction, with steel columns supporting the girders. The columns pass through the buildings, train and subway levels below, and support these various lower levels. The columns rest on spread footings founded on bedrock per the original plans. Much of the existing roadway deck consisted of a reinforced concrete slab with waterproofing, corkboard and a mortar leveling course topped with an asphalt wearing surface. The deck in the remaining areas consisted of a 3/8" thick riveted steel deck plate with waterproofing, corkboard, and a mortar leveling course topped with an asphalt wearing surface. Lichtenstein conducted an in-depth inspection of the Park Avenue Viaduct, including a "hands-on" inspection of all accessible areas of the structure, removal of concrete encasement to confirm member sizes, coring of roadway deck, and a topographic survey of the entire top surface of the viaduct.

The inspection was performed using ladders in the Depew Place section and the viaduct section south of the 41st Street bridge. Lift trucks were used at the 41st, 42nd, and the two 45th Street Bridges. The underside of the deck along Vanderbilt Avenue, areas over the retail rental space from 41st to 42nd Street, in the Helmsley Building and the CONRAIL storage area at the south end of Depew Place was inspected by entering crawl spaces through hatchways.

The in-depth inspection revealed a deteriorated condition of the overlays in all areas and showed signs of heavy leakage at the existing deck joints. Based on this information, it was recommended that the deck be reconstructed and all the expansion joints be replaced with strip seal expansion joints. Special attention would be given to waterproofing and drainage.

In the area where there was an existing reinforced concrete deck, it was recommended that all existing overlays be removed to reduce excessive dead load and upgrade the structure's rating to an H15 (AASHTO TRUCK LOADING). Repairs were recommended to the concrete slab with a new concrete leveling course placed over the existing slab to provide proper cross slopes for drainage, and to adjust the profile grade. A new 1-1/2" latex modified concrete (LMC) overlay would be placed over the leveling course.

In the area where a reinforced concrete structural slab did not exist, it was recommended that all overlay materials and existing waterproofing materials be removed down to the existing 3/8" deckplate. The deckplate would receive new waterproofing and haunches would be created over the floorbeams by placing light weight styrofoam over the deck plate between the floorbeams. A new composite 7-1/2" reinforced concrete slab with a 1-1/2" LMC overlay would then be placed over the deck.

During construction it would be necessary to erect a temporary roof enclosure over the work area in order to protect the rental spaces below from weather elements during the deck removal operations.

The existing drainage system was to be repaired and the existing drainage inlets were to be refurbished.

The ornamental railings and stone balustrades required special attention. These existing details did not meet modern AASHTO Standards for railing design impact loads. The designers held numerous meetings with the Landmarks Preservation Commission and the New York City Department of Transportation Bridge Design Section to develop a protective railing system that would satisfy AASHTO design criteria without compromising the historic character of the ornamental foliate panels. A specially designed two rail system was detailed to be set behind the curb line at a height that would not impair the aesthetic features of the existing railings. The posts were individually spaced to match the existing intermediate supports for the panels. The existing panels were to be removed, cleaned, refurbished, repainted, and reset. Missing or badly damaged panels were to be replaced with matching new sections made from cast iron molds prepared from careful measurements of the existing panels. The granite balustrades were scheduled for special waterblast cleaning, resetting and refitting of granite, and installation of replacement matching sections where needed.

A new lighting system using the existing ornamental

light standards was designed. The existing cast iron standards were slated to be refurbished and fitted with 15Ø watt high pressure sodium lamps. Several replacement posts were to be cast, and this work was to e performed under a separate contract supervised by the New York City Bureau of Electrical Control.

One of the major concerns of this project was the maintenance and protection of traffic during construction. Traffic to the Pan Am Parking Garage had to be maintained and access to the Grand Hyatt Hotel had to be kept open as long as possible. Close coordination with the NYCDOT Bureau of Traffic by the designers resulted in the following maintenance and protection of traffic scheme:

STAGE IA - Southbound roadway closed. Traffic detoured to Fifth Avenue. Reconstruct southbound roadway.

STAGE IB - Closure as above, except northbound roadway closed from 4Øth Street to 42nd Street. Northbound traffic detoured to southbound roadway from 4Øth to 42nd Street. Reconstruct northbound roadway from 4Øth Street to 42nd Street.

FOUR DAY CLOSURE - Necessary to place the LMC overlay from 4Øth Street to 42nd Street, and obtain initial set without traffic induced vibrations.

STAGE IIA - Northbound roadway closed from the Grand Hyatt Hotel to 46th Street. Local traffic for Grand Hyatt Hotel only, traffic exits at Depew Place. Traffic detoured to Third Avenue. Reconstruct closed portion of northbound roadway.

STAGE IIB - Entire northbound roadway closed, detour to Third Avenue. Reconstruct area adjacent to Grand Hyatt Hotel, Grand Central Terminal, and Depew Place Ramp.

In addition, numerous Traffic Control Agents were assigned to the area by the New York City Department of Transportation to aid in the flow of traffic. Lane closures were permitted, but all lanes had to be restored to full use during the period from Thanksgiving to New Year's Day, when the City's Holiday Embargo is in effect. The Embargo prohibits lane closures for construction during this period, and work had to be scheduled accordingly.

Due to the location and extent of work of this project, many agencies and private property owners were involved, and extensive coordination was required in the preparation of final design documents. Management of the overall project and review of bridge design details was done by NYCDOT Bridge Design Section. The maintenance and protection of traffic plans and the location and number of Traffic Control Agents to be used during construction were coordinated with the NYCDOT Bureau of Traffic Operations. Lane closures and scheduling had to be arranged through the NYC - MTCCC - the Mayor's Traffic and Construction Coordinating Council. The redesign of the street lighting system was coordinated with NYC Bureau of Electrical Control and Metro North Commuter Railroad (owners of Grand Central Station).

The final details for the refurbishing of the historic elements were carefully reviewed with the Landmarks Preservation Commission. Mock-up models of railing alternates were placed in the field for review with NYCDOT and Landmarks' representatives prior to the selection of the two-rail system. In addition, methods of construction and scheduling were reviewed in meetings with Metro North Commuter Railroad for Grand Central Station, Grand Hyatt Hotel, Pan Am Building, Helmsley Building, U.S. Post Office, Graybar Realty, New York City Transit Authority, New York Telephone Company, and numerous retail merchants and commercial offices occupying tenant space below the viaduct. It was also necessary to coordinate both our design and construction schedule with the proposed facade reconstruction work at the Pan Am Building.

The proposed construction sequence and staging was prepared to accommodate as many of the conditions requested by the Agencies and Owners as possible, and provided compromises where conflicting interests existed. The final plans reflected a cooperative effort on the part of the City Agencies, Property Owners, and Designers.

The completed design plans were advertised for construction. The low bidder was North Star Construction Co. at approximately $6.5 million.

An extensive series of signs was installed to detour traffic. Advanced warning signs placed well in advance of the project area allowed vehicles to divert before reaching the closure point. This early diversion and the presence of traffic control agents minimized the impacts of the closure on traffic flow. On July 18, 1984, the southbound lane was closed in order to begin reconstruction work (from 40 - 46th Streets).

With the southbound traffic detoured, the Contractor erected the temporary enclosure system over the roadway. The enclosure was framed with structural aluminum sections and covered with polyethylene liner. The roof was pitched for drainage. Completion of the enclosure enabled deck removal operations to begin.

Along the southbound half of the 40th to 42nd Street Viaduct, Deck Removal Operations began with the removal of the asphalt overlay and the underlying mesh reinforced mortar leveling course. Only the structural concrete slab remained. The slab was then scarified. A 2-1/2" minimum concrete leveling course was then placed. Simultaneously, the excavation of the southbound viaduct along Vanderbilt Avenue was in progress. This excavation consisted of removal of all concrete and overlays down to the existing 3/8" steel deckplate. This operation was time consuming, as the work consisted of the removal and trucking away of granite curbs, asphalt block overlay, mesh reinforced concrete, concrete mortar and cork layers. Truck and equipment sizes were limited due to low capacity of the existing structure.

Stud shear connectors wre installed on the floorbeams so that the resulting deck would be composite with the floorbeams providing increased load capacity. The exposed steel deckplate was covered with a 3 ply waterproofing system composed of layers of hot tar and fabric. Styrofoam was placed over the waterproofed deckplate between the floorbeams to create a haunch at the floorbeams, so that the resulting deck would be composite with the floorbeams. At this time bridge rail anchorages were set, epoxy coated rebars were installed and the new 7-1/2" structural slab was placed. Steel face curbing was set with hook anchor bolts drilled into the structural slab at the sidewalk area. Sidewalk reinforcement was installed and concrete was placed. The steel bridge railing was then installed.

Each section of roadway thereafter, involved a different type of construction; for example, for the southbound roadway at the Pan Am Building, the wire mesh reinforced mortar leveling course was removed and the existing 6" reinforced concrete structural slab was to remain. The existing slab was found to be in poor condition, though not predicted by the cores taken during the design phase. An immediate redesign was then performed. The existing slab could not be removed as critical retail rental space existed below in the Pan Am Building and interruption to these operations were not feasible. It was decided to use the existing slab as a form for a new reinforced concrete slab. Pockets were cut in the slab over the floorbeams to provide bearing for the new slab. The existing slab was waterproofed, the rebars were installed, and a new slab placed.

A four (4) day holiday weekend closure from 40th-42nd Streets was arranged. Special signs were erected to detour traffic. A 1-1/2 " latex modified concrete overlay was placed at this time. Wet burlap curing blankets and polyethylene sheets were placed for 24 hours initial set.

The covers were removed, and the LMC was air cured for the remaining 48 hours. The northbound lane was then opened to traffic. During the ensuing work week, LMC was placed on the remainder of the southbound roadway. The southbound roadway was then opened to traffic prior to the Holiday Embargo.

The same operation was done for the northbound lane beginning with its closing on April 1, 1985, and its reopening on September 9, 1985.

The 1986 construction season was spent completing the refurbishing of the railings, panels and balustrades. The existing foliate panels were removed, cleaned, reset and painted. Replacement panels matching the existing were specially cast and installed where needed. The work of rehabilitating the granite balustrades was particularly intricate. The existing sections were cleaned by acid wash. Stone masons reset dislocated segments and placed new matching elements, as needed. The new elements were cut from pink granite matching the existing color and configurations of the original stone balustrade.

The final result, completed in fall of 1986, is a structure capable of supporting modern traffic volumes and loadings while maintaining the historic significance of the structure. The challenge of rehabilitating this critical urban structure under severe traffic and physical constraints while maintaining the historical integrity of the Landmark details, and progressing the project through a number of government agencies is a composite look at the problems facing Civil Engineers in the design of infrastructure repairs.

CONCRETE REPAIRS: ARE WE READY TO DESIGN THEM?

By R. Richard Avent,[1] M. ASCE

Abstract

In-place repair of cracked and deteriorated concrete is frequently performed using one of several available techniques. Epoxy injection, shotcrete, and latex or other modified concrete patching or grouting have been successfully employed. However, the current method of judging the success of a repair is after the fact, i.e., by assessing how the structure performs after the repair. This approach is contrary to established design procedures in which the behavior is predicted through computations prior to initiating the construction phase. In fact, it is customary to use such analyses to evaluate various alternative design approaches so as to select the most effective procedure in terms of strength capacity and economy of labor and materials. At present there are few procedures for predicting the strength after repair for damaged concrete. Design codes are silent on this subject which means that the engineer has no tools to approach the design of a repair scheme in a rational manner.

How close are we to developing the criteria for actually designing concrete repairs? Perhaps closer than most people think. The purpose of this presentation will be to assess progress to date in the development of true design procedures for repair and rehabilitation of concrete structures. First, a review and synthesis of experimental studies will be given. In light of the current state-of-the-art, the primary roadblocks to further progress will be discussed in detail. Finally, some suggested directions and recommendations for developing design procedures will be presented.

While the object of discussion here is deteriorated concrete, the same problem exists in practically every area of infrastructure deterioration. Engineers need design guidance in the form of codes and standards which, at present, do not exist. These needs will be highlighted at the conclusion of the paper with the goal of providing an impetus to future development.

---

[1] Professor of Civil Engineering, Louisiana State University, Baton Rouge, LA 70803

Introduction

Construction methodologies for the repair and rehabilitation of concrete structures have been in practice for at least twenty years. Spearheaded by the needs associated with the nation's deteriorating infrastructure, innovative methods have appeared for repairing various types of damaged concrete. However, these procedures differ in one major respect from that of most construction methods: That is, they are not usually designed in the traditional manner associated with engineered structures. Rather, the usual procedure has been to judge the specific repair method after the fact. An attitude exists, especially among certain speciality contractors, that their repair procedure works for any type of damage. Thus if it worked on the last job, it should work on the next one.

Unfortunately, this same attitude has been adopted by many in the structural engineering profession. Engineers often find themselves in a position of deciding upon a repair methodology without having the tools to effectively analyze and design the repair. There are several reasons why engineers find themselves in such a predicament.

First, the problem of infrastructure repair and rehabilitation is relatively new. Until the 60's the basic philosophy governing repair was to "replace not rennovate." There were sound reasons for this approach. Among the most important were the existance of a highly skilled labor force available at reasonable wages, low interest rates for new construction, and readily available construction materials at low cost. Combined with the American attitude of a throw-away society, the most usual approach to rehabilitation was to tear down and rebuild. However, all of these conditions changed as the nation moved into the 70's. In addition, the structures of the building surge of the thirties were approaching a 40 year life span while structures of the post-world war II building boom were approaching 30 years old. A large inventory of structures have thus now reached or exceeded their expected life span.

Compounding the problem has been the significant gap between the technological level of development in the field of rehabilitation and many other areas of engineering. Most specialized areas of engineering have developed over an extended period of time. For example, the basis for modern structural engineering was established in the last century with new and innovative developments having consistently punctuated progress until the present time. The current state-of-the-art is therefore quite advanced and has been responsible for a "golden age" of structural engineering accomplishments over the past half-century. Since relatively little demand was made on rehabilitation technology, its development has lagged considerably behind other areas of structural engineering. There are relatively few available sources of comprehensive information (how many books can be found on the topic of repair and rehabilitation of structures as compared to steel design, for example?). Excluding those areas that overlap with structural engineering developments for new construction; rehabilitation technology lags significantly, analytical procedures are often primative, and design methodology is often non-existent.

As a result, building codes have discouraged repair and rehabilitation. The building code hierarchy begins within the technical framework of standards and ends with the bureaucratic enforcement procedures at the local level. Complicating the picture is that building code adoption has traditionally been left to each local municipality. The result has been that building standards have varied widely, not only from region to region, but from locality to locality. Combined with the variations in interpretation and enforcement, building codes tend to maintain the status quo and to stifle innovation (8). To combat this tendency, model building codes have been established. These codes tend to be regional with a single code dominating each region. These codes cover all aspects of building construction and when adopted by a municipality, have the effect of law. Within these codes, technical specifications are usually included by reference. Thus the ACI code (1) for concrete construction forms and integral part of practically every building code. These general building codes along with their included technical standards generally have one thing in common: There is practically no mention of infrastructure repair and rehabilitation.

The basic reason for this lack of focus on rehabilitation is that many repair schemes are based on science that crosses the traditional engineering disciplines. For example, the chemistry of the corrosion process, fracture mechanics of concrete, and bond characteristics of adhesives may all play a major role in the design of a repair for a single reinforced concrete beam. The typical structural engineer does not have such a breadth of background nor the time and incentive to build his expertise.

The status just described has been justifiable until recently. Because of the lack of motivating forces to develop repair and rehabilitation procedures in earlier years, the state of the art is just beginning to reach the point of justifying design code inclusion along with the development of analytical procedures, especially for rehabilitation. The purpose of this paper is to assess progress to date in the development of true design procedures for repair and rehabilitation of concrete structures. First, a review and synthesis of experimental studies will be given. In light of the current state-of-the-art, the primary roadblocks to further progress will be discussed in detail. Finally, some suggested directions and recommendations for developing design procedures will be presented.

The Analysis and Design Concept

The concept of structural analysis and design generally incorporates three types of procedures. The first can be classified as the determination of safety factors. In concrete design safety factors are typically provided in the form of load factors. These safety factors depend on the variability of factors associated with the design of a specific component or system. In most cases the safety factors have evolved from long practice with modifications justified by detailed research. The size of the safety factors depends not only on the accuracy of load predictions, but on the accuracy of the other two aspects of design descibed below.

The second category consists of analytical procedures. Design first requires a structural analysis followed by perhaps a more localized stress analysis. These procedures may range from rigorous and formalized general methods (e.g., a linear, elastic analysis of a frame) to approximate estimates and rules-of-thumb (e.g., the Ø factor). In any case, the end result is a set of stresses and/or forces expressing the distribution of loads through the structure.

The final category consists of the determination of basic material properties or the capacity of the basic components. Quite often this type of information is incorporated by reference to specific material standards or specifications.

Each of the categories of information should include standards of design practice for repair and rehabilitation. The essence of the problem is that, with any viable strengthening procedure, there should be an established design methdology which enables the engineer to: (1) Design strengthening schemes, (2) Predict the degree of strengthening for a specific alternative; and (3) Select the most appropriate design from available alternatives. It is in these areas that the most serious deficiencies exist in current rehabilitation technology. In any survey of concrete repair procedures, most conspicuous by its absence is the lack of any evaluation of the analysis/design methodology associated the the various alternatives documented. Once the scope of traditional structural analysis and design is exceeded, there is a dearth of information on computational procedures. As such, practically none of the technical design codes even reference the topic. The practicing engineer facing concrete rehabilitation thus has little guidance as to design considerations. He must rely on his own knowledge and experience, information scattered in diverse technical journals, and his own imagination. While such and approach has resulted in some dramatic successes, it has also led to catastrophic failures. The technology of repair and rehabilitation has now reached a point where it should be included in technical design codes. A concentrated effort in this direction would result in both increased quality in concrete rehabilitation and significant cost savings.

Damage Classification and Methods of Repair

A convenient way to classify concrete damage is by the failure mode. Major failure modes include:

Flexure: Compression zone crushing and tension zone cracks (Fig. 1a, b)

Shear: Diagonal cracking (Fig. 2)

Spalling: Surface pockets

Delamination: Separation surfaces

Rebar Corrosion: Reduced rebar cross section

Rebar Fracture: Effective cross section reduced to zero

Disintegration: Paste/aggregate bond failure

Brittle cracking: Alligator cracking from impact

Each of these damage categories can range from minor to catastrophic. Depending on this level of damage, there are several means available for repair:

Epoxy injection:	Pressure grouting of a low viscosity bonding agent into cracks (Fig. 3)
Patching:	Use of concrete or latex modified concrete to fill surface voids (Fig. 4a,b)
External Steel Reinforcing:	Bonding of steel to external portion of member
Internal Steel Reinforcing:	Drilling and bonding steel to interior sections of member
Addition of external Concrete:	Bonding new concrete to existing (Fig. 5)
Removal and Replacement of Concrete:	Removal of major portions of concrete and replacing with concrete or shotcrete (Fig. 6a, b, c)

A matrix which matches the damage to possible repair procedures is shown in Table 1. In general several options exist for each damage category. Thus the designer needs methods for selecting the appropriate procedure and computing the amount of restoration expected. At present such methods do not exist. However, some data is available to provide a starting point for developing design methodology.

Table 1. Failure Modes and Repair Procedures

Failure Mode	Repair Procedure					
	Epoxy Inject.	Patching	Ext. Steel Reinf.	Int. Steel Reinf.	Ext. Add. Conc.	Replac. Conc. seg.
Flexure	x	x			x	x
Shear	x	x		x	x	x
Spalling		x				
Delamination	x			x		
Rebar Corrosion			x	x		
Rebar Fracture			x	x		
Disintegration		x			x	x
Brittle Cracking	x	x			x	x

A summary of experimental research on repair of reinforced concrete is given in Table 2. The unreferenced tests refer to the author's current ongoing work on repair and rehabilitation. A brief synopsis of each project is described and the projects are grouped and listed in order as to the repair procedure utilized. Two general approaches have been used in these projects to evaluate the effectiveness of the specific repair method. One approach has been to initially load an undamaged specimen to failure, repair the specimen and then re-load to failure. The effectiveness of the repair can be evaluated by comparing the behavior before and after repair. A second approach has been to simulate the damage in the preparation of the specimens, repair the damage and then load to failure. Undamaged control specimens are prepared separately and loaded to failure to provide a means of evaluating the effectiveness of the repair. In either case a good measure of the repair process is the ratio of the restored capacity to the original (or control) capacity. Shown in Table 2, as a percentage, are the results of the reported tests based on the average level of the restoration. This figure not only provides insight as to the effectiveness of the repair, but also may provide guidance on developing analysis/design procedures for predicting the strength after repair. In each of the following sections, concrete repair methods are evaluated in terms of the available test results. A guide as to further research needs is also presented.

Epoxy Injection.--As illustrated in Table 2, two general types of tests have been conducted to evaluate the effectiveness of epoxy injection: (1) small plain concrete specimens such as cylinders or blocks; and (2) reinforced concrete beams or beam assemblies. The first three entries in Table 2 evaluate the compressive strength of epoxy injected cylinders. Notice that there is considerable variation among the test results. However, the trend shows that epoxy injection can restore compression strength in the range of 90 to 110% of undamaged cylinder values. The high strength results reported by Avent and Kongsuwan (2) point out the danger of using small specimen tests exclusively. The cylinders tested were 3 x 6 inches (8 x 15 cm) and were heavily damaged with the initial loading. The amount of epoxy required to repair them resulted in circumferential restraint hoop stresses which tended to magnify compressive strength. In addition, the large amount of epoxy compared to concrete was uncharacteristic of typical applications and distorted the results.

Less than 100 cylinder tests have been reported so the sample size is too small for developing design related criteria. The basic need is to conduct a comprehensive series of compressive strength cylinder tests emphasizing the effect of initial damage level, variations in 28 day strength, and stress-strain characteristics of the material. The availability of this data may serve as a guide to developing a modified stress block for use in computing the ultimate strength of epoxy injected members or, in developing adjustment factors such as a modified Ø factor.

The additional small specimen tests shown in Table 2 relate to the strength of plain concrete in flexural tension and shear. Typically, epoxy injection provides a 10% increase in these strength factors. The failure mechanism for repaired specimens is usually similar to the original failure with the location being shifted away from the glue line. These tests provide a good indication of the bond strength of epoxy to concrete. In general it is equal to or greater than the corresponding strength of the concrete

Table 2. Summary of Test Results for Various Concrete Repair Procedures.

Author	Specimen Type	Number of Specimens	Test	Repair	Average Recovery Strength	Comments
Avent & KOngsuwan (2)	3" x 6" cylinders	10	Compression	Epoxy injection	171%	Slight decrease in stiffness
Fuller & Kriegh (9)	6" x 12" cylinders	15 20 21	Compression Tension Combination	Epoxy injection	89%	
Wakeman, Stover & Blye (14)	2" x 10" cylinder cores	9	Compression	Epoxy injection prior to coring	110%	Damage was repaired at site and cores taken
	6" x 6" x 24" unreinforced beams	9	Simple Flexure	Epoxy injection	107%	Epoxy bond line did not reopen
Chung (5)	8" x 16" ell-shaped concrete specimens	4	Static shear	Epoxy injection	109%	
Chung & Lui (6)	8" x 16" ell-shaped concrete specimens	8	Impact shear	Epoxy injection	112%	
Chung (4)	5" x 8" x 9' simple span R/C beams	3	3rd point loading to ultimate	Epoxy injection	104%	Slight increase in stiffness, sealed cracks did no reopen
Fuller & Kriegh (9)	Simple span R/C beam	1	Flexure	Epoxy injection	95%	Stiffness reduced
Popov & Bertero (13)	Beam-column sub-assemblies with moment resisting connections	3	Initial monotonic loading followed by cyclic loading after repair	Epoxy injection with grout of spalled areas	100%	Hysteretic loops narrowed and pinched, stiffness reduced
Avent & Kongsuwan (2)	8" x 14½" x 18' simple span R/C beams	11	3rd point loading cyclic & monotonic	Epoxy of spalled areas with epoxy injection	95%	
Author	4.5" x 9.5" x 7.75' R/C simple span beams	6	3rd point flexure	Grout of spalls with latex modified concrete	118%	Comparisons made to control specimens
		6	3rd point flexure	Grout of spalls with concrete and epoxy at bond line	95%	
McDonald & Calder (12)	5.9" x 9.8" x 11.5' simple span unreinforced beams	19	3rd point loading	Steel plate epoxy bonded to concrete	173%	Ductile failures could be obtained with proper proportions

Table 2. (Continued)

Author	Specimen Type	Number of Specimens	Test	Repair	Average Recovery Strength	Comments
Holman & Cook (11)	4" x 8" R/C beam	9	Torsion	Steel plate epoxy bonded to concrete	133%	Stiffness was increased and failure mode was ductile
Chung (5)	8" x 16" ell-shaped concrete specimens	4	Monotonic static load producing surface shear	Epoxy injection with steel reinforcing		
Chung & Lu (6)	8" x 16" ell-shaped concrete specimens	8	Impact load on shear plane	Epoxy injection with steel reinforcing		
Chung (3)	concrete cubes, R/C beams	9 2	Rebar pull-out	Epoxy injection to bond rebar		
Cowell, Popov & Bertero (7)	R/C block cast around steel bar	2	Cyclic bar pull-out	Epoxy injection		
Geymayer (10)	9" x 4" x 6.5' R/C beam	14	3rd point flexure loading	External addition of 1.5 to 3 inches of epoxy/polyester to tension face of beam	None Reported	Comparative study of resin mixtures
Author	4.5" x 9.5" x 7.75" R/C simple span beams	2	3rd point flexure	External addition of modified latex concrete to compression zone	117%	Comparisons made to control specimens
		2	3rd point flexure	External addition of concrete grout to compression zone with epoxy at bond line	94%	
Author	4.5" x 9.5" x 7.75" R/C simple span beams	4	3rd point flexure	Replacement of entire segment with latex modified concrete	106%	Comparisons made to control specimens
		4	Midpoint flexure	Replacement of entire segment with concrete and epoxy at bond line	102%	
		4	3rd point flexure	Replacement of entire segment after shear failure with concrete of epoxy at bond line	106%	Comparisons made between initial failure and failure after repair

itself in either tension or shear. While additional testing may be helpful, research emphasis needs to be directed toward member behavior.

Table 2 has four entries describing epoxy repair of reinforced concrete beams. The trend of these tests indicates 100% restoration can be obtained. However, the sparseness of the data (18 beams) precludes direct incorporation into design applications. A number of variables need to be studied in more depth. Principal among these are: (1) level of damage (e. g., some tests show that the restoration level is inversely proportional to the degree of damage); (2) Failure type (e. g., Effectiveness varies depending on whether the damage is crushing, shear cracking, rebar slipage, or tension cracking); (3) Loading conditions (particularly monotonic versus cyclic); (4) stiffness after repair (some tests indicate a reduction in flexural stiffness which may influence analytical procedures); and (5) Strength and age of the concrete. A study of individual tests indicates considerable variation which may be due to any of the above factors. A careful documentation of additional studies should lead to criteria useful for predicting restoration strength. It appears likely that for some cases the restoration level will be lower than 100% while for other situations the level may be as high as 130%. Based on available test results, a typical restoration level for epoxy injected flexural members should probably be taken as 90%.

Patching with Concrete Grout.--An effective approach for repairing spalls or concentrated pockets of damage is to remove the damaged portion of concrete and replace with a concrete grout. Several grouting procedures are used with the most popular being to use a concrete mix with additives such as latex to provide increased adhesion on the bond surface. An ordinary concrete mix can also be used but a film of epoxy should be placed along the bond surface and the concrete poured before the epoxy sets. Recent test results by the author as shown in Table 2 indicate that latex modified concrete can provide a significant strength increase. The use of ordinary concrete with epoxy at the bond surface is not as effective but the level of restoration is about 95%. Considerable testing needs to be conducted considering the same variables as listed for the epoxy injection method. Additional factors to be considered are the relative strength of the grout and the original concrete along with the shrinkage characteristics of various grouts.

External Reinforcement.--The use of external reinforcement such as steel plates bonded or clamped to the concrete offers a promising method for effective repairs, especially if the desired restoration level is to exceed the original strength by a significant margin. The test results reported in Table 2 show that strength levels can be increased by 33 to 75% over original strength. Again the small number of specimens tested to date make it impractical to formulate design procedures. However, this approach perhaps offers the greatest potential for repair of concrete structures. A number of variations involving external steel in conjunction with epoxy injection and grouting offer great potential. Needed now are additional studies on the important parameters influencing the design strength of such arrangements. Among the more important parameters are: bond characteristics, size and properties of the external reinforcement, effective arrangements for various types of damage, corrosion protection, and analytical processes for designing the repair. At present, research data are so limited that applications require extensive testing which is usually not practical for the practicing engineer.

Internal Reinforcement.--Table 2 illustrates the results of several studies related to internal reinforcement. The typical process is to insert reinforcing into either existing holes or, into drilled holes, and bonding with epoxy grout. A similar application is to re-bond existing bars that have slipped. The insertion of new bars has been shown to be particularly effective. Strength increases as high as 45% above original have been obtained. Re-bonding of existing steel has not been as successful due primarily to the difficulty in getting epoxy to penetrate the entire depth of the reinforcing rod.

The use of internal reinforcement illustrates the need for development of design criteria. The engineer needs to be able to calculate the size, location, spacing, and length required for a given level of restoration. While methods associated with the design of new beams may provide an estimate, many repair applications simply do not fit the pattern and behave differently. At present the engineer is on his own when attempting to design such a system. An analytical and experimental program is needed to provide these tools to the practicing engineer.

Addition of Concrete to External Surface.--The use of this approach is an extension of the grouting procedure for spalls and can increase the strength of the member above its original. Results in Table 2 indicate that strength levels can range from 95 to 117% although the small number of tests only suggests the potential for this procedure. Careful testing and evaluation is needed to quantify design parameters as described in previous sections. Additional concerns are related to shrinkage problems and the compatibility of different strengths for old and new concrete.

Replacement of Entire Segment.--Sectional replacement using shotcrete or cast-in-place grout has become quite popular. However, relatively little testing has been conducted. The assumption has been made that existing design procedures for new members will apply equally to this repair scheme. The test results shown in Table 2 indicate that the method is effective. However, the samples tested form a small data base. Interaction of old and new concrete, strength variations, and other concerns still need to be answered.

Future Directions

A basis had been laid for developing design concepts for repair of reinforced concrete structures. However, at present design is limited to carrying over the applicable concepts associated with ordinary design of new members. Such an approach tends to stifle innovation and limit applications. Code incorporation of repair design concepts is also precluded by this approach. The success of progress will depend on several factors.

First, researchers must recognize the need for sound analytical development in the area of repair and rehabilitation. Repair does not have the glamor associated with new analysis concepts. Research also tends to cross discipline lines making progress more difficult. As such there are few courses, or texts describing the state-of-the-art related to repair and rehabilitation. The learning curve for new researchers is thus longer and often more difficult than for many other areas of structural engineering.

Second, the profession must recognize the need for development of the design process. Too many practicioners are satisfied with the status quo. "If it works don't fix it" reflects this attitude. If such an attitude had prevailed in the past, we would still be designing concrete structures using the "working stress" method.

Third, code writing bodies need to address the issue of including aspects of repair and rehabilitation in design codes. Widespread acceptance will not result until codes provide a framework for standards of practice.

Finally, more research emphasis should be placed on developing design concepts by research funding organizations. The most notable progress in this area perhaps lies with those organizations tied to federal highway problems such as the Federal Highway Administration. However, even with such examples, most agencies are far too demanding for immediate results. Truly innovative development in the area of concrete repair and rehabilitation will require some basic research. This aspect should be recognized and a portion of funding earmarked for such purposes. Far too often, so called "research projects" are simply a compilation of previous research with no real development. However, a prime illustration of the positive approach is the development of earthquake engineering during the last several decades which can be related directly to the emphasis of funding organizations such as NSF.

Appendix.--References

1. American Concrete Institute, Building Code Requirements for Reinforced Concrete (ACI 318-77), ACI, Detroit, MI, 1977

2. Avent, R. R., and Kongsuwan, S., "Behavior of Epoxy Repaired Concrete Bridge Girders," Proceedings, Bridge Maintenance and Rehabilitation Conference, Department of Civil Engineering, West Virginia University, August, 1980

3. Chung, H. W., "Epoxy Repair of Bond in Reinforced Concrete Members," ACI Journal, Vol. 78, No. 7, Jan/Feb, 1981

4. Chung, H. W., "Epoxy-Repaired Reinforced Concrete Beams," ACI Journal, Vol. 72, No. 5, May, 1975

5. Chung, H. W., and Lui, L. M., "Epoxy Repaired Concrete Joints," ACI Journal, Vol. 74, No. 6, June, 1977

6. Chung, H. W., and Lui, L. M., "Epoxy-Repaired Concrete Joints Under dynamic Loads," ACI Journal, July, 1978

7. Cowel, A. D., Popov, E. P., and Bertero, V., "Repair of Bond in R/C Structures by Epoxy Injection," Proceedings, U.S./Japan cooperative Research Program in Earthquake Engineering on Repair and Retrofit of Structures, Dept. of Civil Engineering, Univ. of Michigan, Ann Arbor, May, 1980

8. Field C. G., and Rivkin, S. R., The Building Code Burden, D. C. Heath and Co., Lexington, Mass., 1975

9. Fuller, J. D., and Kreigh, J. D., Maintenance and Repair of concrete and Masonry Structures: Epoxy Grouting, Dept. of the Army Construction Engineering Research Laboratory, Technical Report M-9, July, 1971

10. Geymayer, H. G., Use of Epoxy or Polyester Resin Concrete in Tensile Zone of Composite Concrete Beams, Technical Report C-69-4, Waterways Experiment Station, Corps of Engineers, Vicksburg, MS, Mar, 1969

11. Holman, A. M., and Cook, J. P, "Steel Plates for Torsion Repair of Concrete Beams," Journal of Structural Engineering, ASCE, Vol. 110, No. 1, Jan, 1984, pp. 10-18

12. McDonald, M. D., and Calder, A. J. J., "Bonded Steel Plating for Strengthening Concrete Structures," International Journal of Adhesion and Adhesives, April, 1982, pp. 119-127

13. Popov, E. P., and Bertero, V., "Repaired R/C Members Under Cyclic Loading," Earthquake Engineering and Structural Dynamics, Vol. 4, 1975, pp. 129-144

14. Wakeman, C. M., Stover, H. E., and Blye, E. N., "Glue! For Concrete Repair," Materials Research and Standards, Vol. 2, No. 2, Feb. 1962

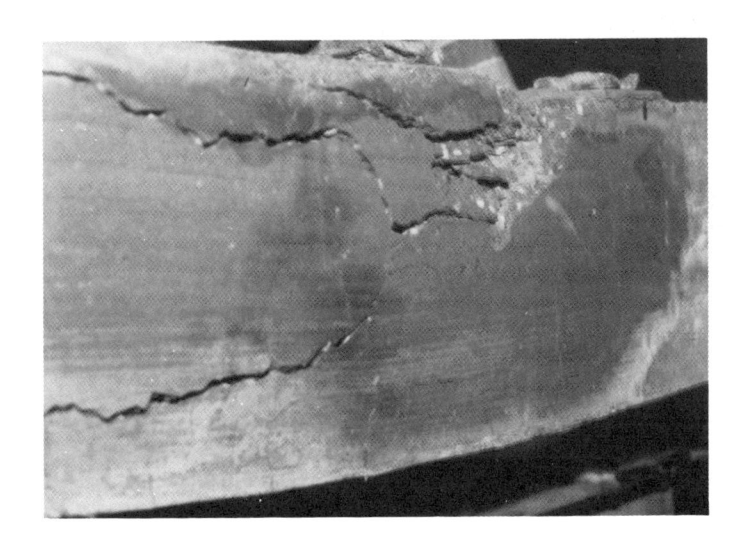

Figure 1a. Compression zone crushing of an R/C flexural member

Figure 1b. Typical cracks in an R/C flexural member

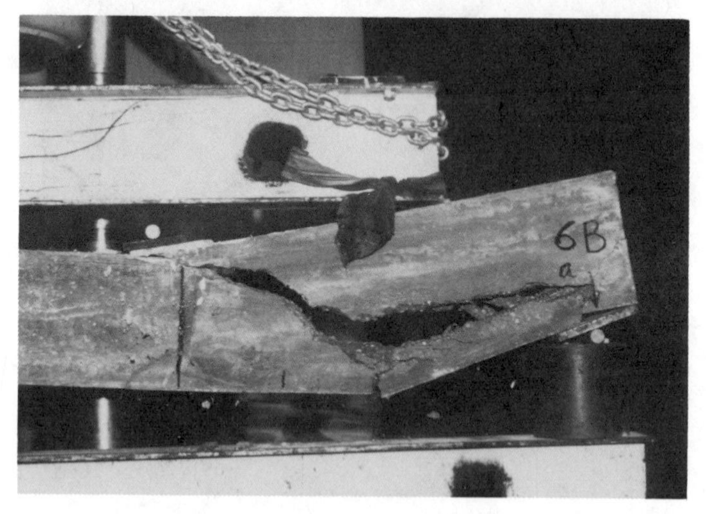

Figure 2 . Diagonal cracking associated with a shear failure

Figure 3. Sealed and epoxy injected reinforced concrete tee beam

Figure 4a. Example of patching voids with concrete grout in a flexural member (The darkened area to the left is the original repair while the damage on the right resulted from re-testing the beam)

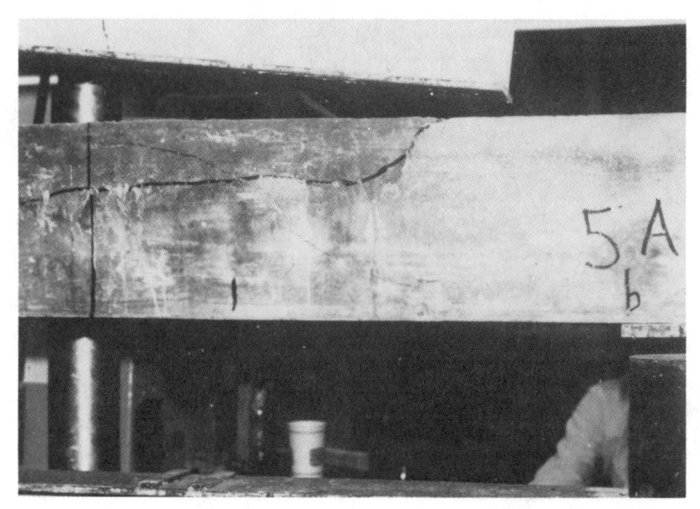

Figure 4b. Example of latex modified concrete grout for patching voids (The grouted area is the upper third of the beam on the left portion of the photograph)

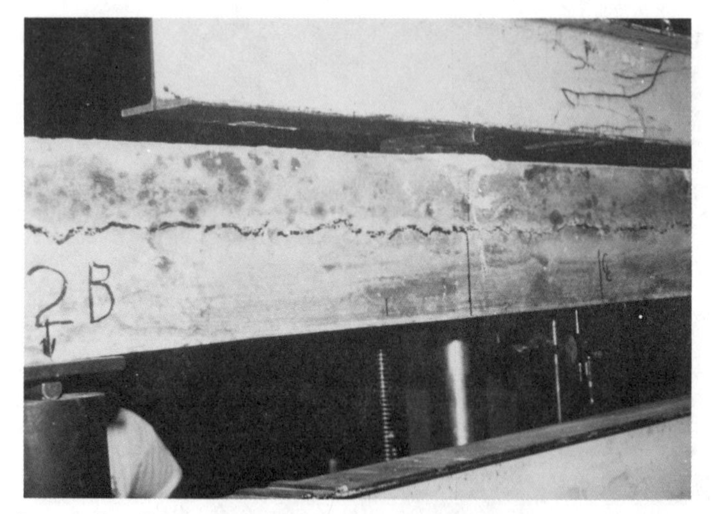

Figure 5. Example of bonding new concrete to an existing beam (The new concrete has been added to the top of the beam along its entire length)

Figure 6a. Example of repair of the full depth of a reinforced concrete beam with new concrete (the dark area indicates the repair zone and the damage is due to a second test after the repair)

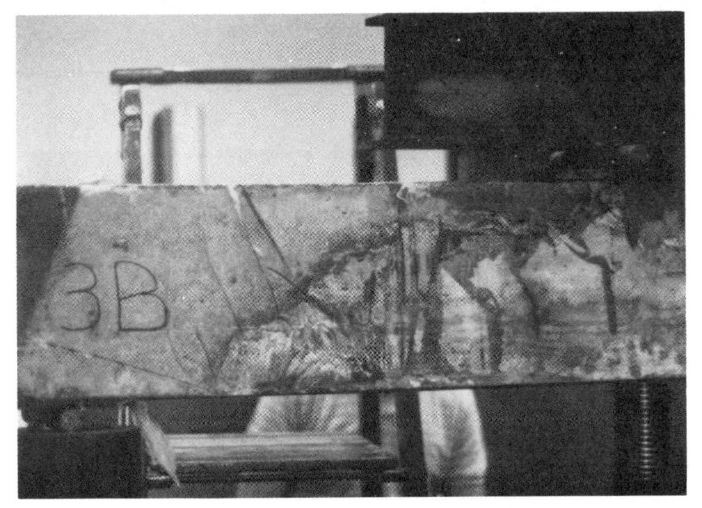

Figure 6b. The left hand section of this beam has been repaired by adding full-depth concrete bonded to the existing beam with epoxy at the bond surface

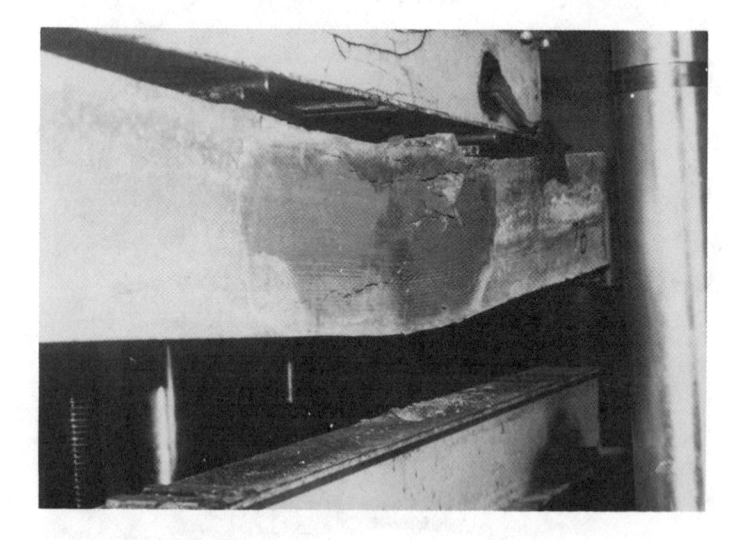

Figure 6c. The entire middle portion of this beam was repaired with latex modified concrete (The damage resulted from a re-testing of the beam)

Concrete Bridge Deck Condition Assessment:
Traditional and Innovative Inspection Technologies

W. M. Kim Roddis,
Associate Member ASCE*

1. Abstract

Bridge deck deterioration is a major problem for highway agencies and is one of the leading contributors to the number of deficient bridges in the United States. A reliable assessment of the condition of a bridge deck must be available if a cost effective and systematic approach is to be taken for deck rehabilitation. The frequency and depth of deck inspections can be tailored to suit bridge management needs by appropriate utilization of the various available inspection techniques. Traditional inspection techniques are described and the capabilities and limitations of each method are presented. The shortcomings of traditional methods, especially in the case of asphalt overlaid decks, are summarized and the desired performance criteria for alternative methods are delineated. Nontraditional testing methods are critiqued for potential application to deck condition assessment. The two most promising technologies are ground penetrating radar and infrared thermography. Capabilities and limitations of these methods are discussed, both as inherently defined by the measured phenomenon, and as determined by the current maturity of the application technology. Issues requiring resolution for routine application of radar and thermography in routine deck assessment surveys are outlined.

2. Statement of the Problem

Bridge deck deterioration is a major problem for highway agencies and is one of the leading contributors to the number of deficient bridges in the United States. For this reason, significant bridge rehabilitation efforts have focused on correcting deck deterioration. In spite of these efforts, deterioration of bridge decks continues to take place at a faster rate than repairs are made.

Rational bridge management decisions cannot be made without sufficiently accurate, detailed and reliable information on deck condition. There are many places in the allocation of bridge maintenance resources where information on deck condition can be used to improve the quality of decisions. These areas include: inspection, preventive maintenance, priority of projects, preparation of bid documents, construction quality, evaluation of the performance of repairs, and prediction of future needs. A reliable assessment of the condition of a bridge deck must be available if a cost effective and systematic approach is to be taken for deck rehabilitation. Defects and deterioration must be identified in order to establish priorities for rehabilitation and to decide on the appropriate method of repair or

---

*Fannie and John Hertz Foundation Fellow, Civil Engineering Department, MIT, Cambridge, MA, 02146.

replacement.

Accurate determination of deck condition is complicated by the fact that signs of degradation are not usually visible at the surface until the deterioration is far advanced. The primary cause of deterioration is corrosion of the reinforcing steel, a mechanism which has been accelerated and amplified in recent decades by the widespread use of deicing chemicals. Reinforcement corrosion deterioration is a subsurface problem so the progress of the damage is hidden until the deck actually begins to spall. The difficulty of detecting damage before it is far advanced is compounded if the concrete deck is hidden from view by a bituminous overlay.

3. Current inspection methods

The various well-established methods of data collection for deck condition appraisal may be roughly ordered from most to least frequently used as: visual inspection, delamination detection by sounding, measurement of chloride content, core drilling and testing, measurement of corrosion potentials, cover measurement by pachometer surveys, and electrical resistance testing of membrane integrity.

3.1. Visual inspection

The first step in any condition survey is usually a visual inspection. All visible defects on both top and bottom deck surfaces are categorized and recorded. The size, location, extent, and depth of spalling and scaling are noted. The location, length, width, and orientation of cracks are observed, and if possible, the cause of cracking is determined. Visual inspection is complicated if an asphalt overlay hides the deck. Overlay cracking and wet spots may be indicative of underlying deck distress. If the underside of the deck is accessible, it is examined for signs of leakage and deterioration such as: efflorescence, wet spots, cracking, and rust stains [Park 80].

Visual inspection, when done by an experienced bridge inspector, has been found to be the single most reliable and informative source of information for condition assessment. However, the reliability of visual inspections depend on the perceptiveness and judgment of the inspector, and there are not always enough experienced inspectors to do the job. The supply of trained inspectors varies with location, so some areas have difficulty finding enough qualified inspectors while others do not.

Visual inspection, although an extremely useful method for bare decks, is a problematic method when applied to asphalt-covered decks, even when performed by experienced inspectors. The condition of the bituminous overlay may or may not be indicative of the condition of the underlying concrete deck. In cases where an effective membrane is in place, the overlay may be in poor shape when the deck is in good condition. Conversely, when no membrane or an ineffective membrane is present, the deck may be severely deteriorated when the overlay is in good condition [NCHRPS118 85]. The top visual inspection is not very informative, and the bottom visual inspection can only identify areas of extreme deterioration. If the underside of the deck is not accessible for inspection, or is hidden by stay-in-place forms, and an overlay is present, visual inspection is not of much use.

3.2. Detection of delaminations by sound

Delaminations mechanically separate the upper layer of the concrete from the bulk of the deck. When a delaminated area of the deck is struck with a tool, this separation makes the noise sound dull and hollow in comparison to the more highly pitched ringing noise of an intact portion of deck. Hammers, rods and chains are used to strike the deck. Chains are most common, so this type of test is sometimes simply called a "chain drag" test. This traditional method of identifying delaminations has been found to be effective for inspection of exposed concrete decks [Manning 82].

Sounding techniques, such as chain drag, accurately detect delamination on bare decks. This approach is inexpensive and not weather dependent but is tedious, dependent on the operator's skill, and time consuming. It is difficult to use on a bridge only partly closed to traffic, due to interference from traffic noise. Sounding methods have difficulty in differentiating between debonding and delamination. In addition, chain drag identifies only a small fraction of the delaminations on asphalt-covered decks [Manning 82].

Efforts have been made to automate this delamination detection process in order to increase its speed, decrease its dependence on operator skill, and decrease the tedium of recording data. A commercial product, the Delamtect [NCHRPS57 79], consists of a tapping device, a sonic detector, and a pen recorder, all mounted on a small hand pushed cart. Since the Delamtect only tests a 150mm (6") wide strip, a series of passes in a grid pattern are necessary to obtain adequate deck coverage and small isolated delaminations would be expected to be missed at the 1.5 m (4') grid spacing commonly used [Manning 82]. This device is not as accurate as hand methods. Accuracy of this instrument degrades substantially when it is used on asphalt-covered decks [Manning 82]. For these reasons, manual and automated sounding techniques are not promising for use on asphalt-covered decks.

3.3. Chloride content chemical testing

Chemical analysis of a sample of concrete from the level of the top reinforcing mat can be used to determine whether the deck is contaminated with chlorides. The number of samples that must be taken to obtain a data set representative of the overall deck condition is dependent on that deck's variation in chloride content. Six samples are commonly recommended as a minimum set. Different approaches are taken to locate samples. Samples can be randomly located, located in an attempt to maximize variation, or located in areas of ambiguous results from other test methods. Samples may be taken in either cored or pulverized form for laboratory analysis. A rapid in situ method has also been developed, which is less accurate than laboratory methods, but has the advantages of being quick and causing minimal, easily repaired deck damage [NCHRPS57 79].

3.4. Core drilling and testing

Another sampling technique that is commonly used to supplement nondestructive methods is core drilling. Since this is a discrete sample method, questions of sample number and location again arise. On exposed decks one core is usually taken for every 2000 square feet of deck, with a minimum of three cores. For an asphalt overlaid deck this amount would be increased three or four fold. In this case, sections of the overlay are usually removed in rectangular patches to allow visual inspection of the deck at sample locations. Cores are usually taken in these stripped areas. The cores are examined in the laboratory to determine: visual signs of deterioration, aggregate and cement paste condition, air voids, density, strength, and chloride content.

Coring and chloride ion content tests may give results that are not representative of the overall decks, since a limited number of samples are taken. Sampling methods of coring and chloride content can give extremely misleading results if the samples are taken from a part of the deck that the person collecting samples thinks are typical locations which are actually non-representative parts of the deck, for example over beams. In addition, the significance of a particular chloride content is ambiguous since it is not total chloride but only free chloride which is available to destroy passivation. Since the correlation between free and combined chloride is not well defined, inferences based on the measured total chloride may be misleading.

3.5. Corrosion potentials

As the reinforcing rusts, current flows within the macrogalvanic cell from the anode to the cathode. This current flow creates a potential difference between the anodic half cell and the cathodic half cell. A voltmeter can be used to measure this corrosion potential against a reference potential. The standard version of this test uses a copper/copper-sulfate (CSE) cell to provide the reference voltage. The standard equipment and procedure are specified by an American National Standard [ASTM 85]. A reading of less than -0.20 volts CSE is interpreted as greater than 90% probability of no active corrosion in the area; the range from -0.20 to -0.35 volts CSE is uncertain; and a reading of more than -0.35 volts CSE is interpreted as greater than 90% probability of ongoing corrosion. This method detects only the likelihood of corrosion and does not detect the rate of corrosion. When used on asphalt-covered decks, it is desirable, if no membrane is present, and essential if a membrane is present, to drill through the paving to ensure contact with the concrete [Manning 82].

Some users have found the corrosion potential test, when used in a grid pattern, to be a very useful indicator of corrosion activity. Others have had poor success with this method. It has the disadvantages of requiring lane closures which are may be unacceptable for dense urban areas.In addition, for bridges which have a membrane, it is necessary to drill through the asphalt and puncture the membrane at the grid points, which would be detrimental if the membrane was previously intact. If the deck is paved, large amounts of moisture in the paving can conduct electricity over a broad area so the area of corrosion activity may not be identified. The test cannot be performed when the deck is frozen and it is recommended that both deck and ambient temperature be above 50 degrees F.

3.6. Pachometer surveys of cover depth

The thickness of the concrete cover over the top layer of reinforcement can be measured by a magnetic device called a pachometer. The pachometer generates a magnetic field between two poles on a probe. This field is distorted by ferromagnetic materials, such as steel. The pachometer detects the magnitude of this field distortion, which is proportional to the size of the bar and its distance from the probe. If the size of the bar is known from construction drawings, the cover thickness can be determined directly, otherwise cores can be used to determine bar size at several points. Cover measurements are used to: establish the depth of the top steel for taking chloride samples, determine if low cover is the cause of observed deterioration, and locate areas with insufficient cover to allow scarifying [Park 80].

Pachometer surveys of cover depth do not provide direct information about deterioration, but instead identify areas of low cover which are likely to experience distress. If an overlay is in place, the pachometer identifies the distance from probe to bar through both overlay and over, not the desired measure of concrete cover.

3.7. Electrical resistance of membranes

Most membrane materials are not electrically conductive. The electrical resistance of a membrane can be assumed to be indicative of its permeability since the openings that allow moisture to pass will also allow current flow. The resistance of the membrane can be measured by connection of an ohmmeter to the top of the membrane and the top layer of reinforcing. The standard equipment and procedure to perform this test are specified by an American National Standard [ASTM 85]. Since epoxy is an insulator, decks with epoxy coated top mats cannot be tested with this method. In addition, pavement porosity and moisture variability can cause problems in obtaining meaningful readings. If the readings are taken from the paving surface instead of stripping the overlay at the test points, the readings are not very reliable. This test method is susceptible to error. There is no general agreement on interpretation of the data so the evaluation of the results is subjective. Electrical resistivity tests are unable to discriminate between membranes with a few pin holes, which offer good deck protection, and those with large punctures, which fail to offer protection [Manning 86].

3.8. Shortcomings of current methods

The major limitations of all current methods are that they are slow, labor intensive, and require lane closures. Consequently, they can only be applied to a limited number of decks, usually those in the worst condition. Visual inspection of sample areas which have been stripped of their bituminous overlays; petrographic examination of cored specimens; chloride content; and potential tests all share the drawback of either depending on a small number of samples and hence risk being non-representative of the overall deck or requiring an extensive grid of test points and hence become time consuming and labor intensive. In addition, the lane closures required for these tests are unacceptable in urban areas. The fact that methods like visual inspection require a high degree experience in order to be reliable and complete is an obvious weakness. In summary, in spite of the effort to systematically apply traditional methods, it is very difficult to assess the condition of an overlaid deck with the accuracy needed for maintenance management subject to tight funding constraints.

4. Desiderata for new methods

The need for improved non-destructive testing methods for bridge deck evaluation has been clearly established. A wide variety of testing methods are available for consideration [Bungey 82]. In order to identify the most promising methods, the desired characteristics of a test method need to be made explicit. The performance of candidate methods can then be compared against the desired criteria so that the best methods may be identified.

A practical testing system should have the following characteristics:

- Nondestructive, so the condition of the in-service deck is not degraded and so the test may be repeated at a later time.

- Rapid, preferably noncontact, so the extent of deck surveyed may be large and so traffic is not constricted.

- Access independent, so testing is done only from on top of the deck.

- Reliable and accurate, to obtain information adequate for the decision process.

- Economical, to ensure cost-effectiveness and allow broad coverage.

- Flexible, so it may be applied to general deck configurations, including overlaid decks where the inadequacies of traditional methods are greatest.

- Weather tolerant, so testing may be performed under a range of environmental conditions.

- Rugged, so the equipment is suited to field use.

- Objective interpretation, so results are repeatable, quantifiable, and not dependent on the availability of skilled operators.

It is also desirable for the techniques to be relatively technologically mature. This would allow rapid field use since the development of new types of test equipment would be avoided and instead the most promising existing equipment would be used to develop applications and procedures to produce the condition data needed to make bridge management decisions.

5. Summary of nontraditional methods considered

5.1. Rebound and penetration tests

Rebound and penetration methods use a hardness measure to indirectly predict concrete strength. Several existing tests are: indenter tests, Schmidt hammer, and Windsor probe. All these methods require exposed concrete for their use and none are reliable for doing more than identifying anomalous areas without calibration for each deck by some other test [NCHRPS118 85]. These types of test method were not further investigated since all methods based on the indirect inference of concrete strength from surface hardness would have these drawbacks and therefore be of marginal interest for deck assessment.

5.2. Sonics and ultrasonics

Sonic and ultrasonic techniques use sound waves. The presence of cracks, reinforcement, voids, microcracks, and moisture affect the transmission of sound in concrete so these methods have been investigated for applicability to deck assessment. Sonic and ultrasonic pulse-velocity methods use a velocity measure of mechanical stress waves to infer concrete strength. These methods are indirect (the velocity is related to the elastic modulus which is related to the void content which is related to the compressive strength) since the material property of interest is only indirectly measured by the test energy. These methods have the undesirable characteristic of requiring mechanical contact for both source signal and measurement.

Sonic reflection techniques are based on monitoring the audible sound produced by striking the deck. The presence of delaminations is detected by a change in frequency of the sound of an impact on the deck. The traditional chain drag test and the automated Delamtect device are sonic techniques that have been investigated for use on asphalt-covered decks. Sonic reflection has been found to have a very low accuracy for this application [Manning 82].

Ultrasonic transmission involves introducing high frequency sound waves into the deck and measuring the time of wave travel. The speed of sound in a homogeneous material is a function of the material's density and elastic constants. In concrete, the heterogeneous composition, porosity and moisture content also affect the speed of sound. The reflections of the sound waves at interfaces can be used to detect discontinuities. Ultrasonic transmission has certain limitations which render it impractical for concrete deck surveys [Manning 82]. If direct transmission of pulses is used, the source and receiver must be located on opposite sides of the deck with a known path length between the transducers. The path length through a deck is not known with sufficient accuracy to detect signal changes due to variation in concrete quality. Indirect transmission, with both transducers located on top of the deck, has even less path length certainty and hence cannot produce meaningful results.

5.3. Seismic and microseismic waves

Seismic waves are capable of resolving only large discontinuities and therefore are not applicable to deck assessment. Their main application area is to geophysical investigations. Microseismic refraction uses spaced geophones to time the travel of a shock wave through the test material. When investigated for bridge decks [Manning 82], microseismic refraction detected defects, but could not define their extent. The method is slow and difficult to interpret.

Another study [Murphy 85] measured the velocities of seismic compression waves in a thick bridge deck. The velocities were indicative of the concrete compressive strength based on an empirical relationship. Like sonic and ultrasonic pulse-velocity methods, this seimic method is indirect and requires mechanical contact.

5.4. Acoustic emissions

Acoustic emissions are the low (often in the 50 to 100 KHz range) frequency sounds that most materials emit as they are deformed. As a material is loaded, kinetic energy is released by localized yielding, crushing, or microcracking. This energy release produces small amplitude stress waves which propagate through the material [Bungey 82]. Although acoustic emission techniques have been cited as having potential for bridge deck evaluation, the applications to concrete are not fully developed and are still regarded as essentially laboratory methods. Serious technical difficulties need to be surmounted before acoustic emission can be used as a field testing method [NCHRPS118 85].

5.5. X-ray and gamma ray sources

X-rays and gamma rays are distinguished only by their origin. X-rays are produced by extra-nuclear atomic processes while gamma rays are usually nuclear in origin [NCHRPS118 85]. Local density gauges using gamma rays and density mapping techniques using x or gamma rays could detect deck defects [Joyce 84]. Nuclear density gauges using an uncollimated gamma source can make rapid average density measures for the volume of deck near the gauge, but cannot measure local density variations. A Compton scatter gauge, measuring the collimated back scatter of a collimated gamma source, is sensitive to local density variations, but has a limited field of view and requires long time, on the order of several minutes for each data point.

Computerized tomography is a density mapping technique that offers the attractive concept of developing a nondestructive cross-sectional view of the deck. For a stationary object, tomography obtains a clear image of one plane, while blurring all other planes, by moving the radiographic source and film. Commercial equipment is not yet available and the practical feasibility of this method has not been established [Joyce 84]. More conventional techniques based on x-rays are not sensitive to fracture planes perpendicular to the direction of radiation [Manning 82] and require access to both sides of the test sample.

Methods based on x-ray or gamma ray sources do not appear to be easily applicable to deck assessment and only show significant promise as an inspection technique for prestressed concrete structures [NCHRPS118 85].

5.6. Optical methods

Optical techniques have the advantage of being rapid and high resolution. Direct visual observation of top and bottom slab surfaces is part of current inspection practice. Laser and other optical methods do not appear to have promise for detection of defects in asphalt overlaid decks, due to their disadvantages of being only a surface test and requiring a line of sight.

5.7. Electric methods

Resistivity/conductivity measurements are used to infer the integrity of membranes and sealers. Potential tests indicate the likelihood of active corrosion, but do not provide information on the corrosion rate. Resistivity and potential tests are elements of current practice for deck assessment.

Alternating current impedance tests can be used to estimate corrosion rates [John 81]. Since impedance methods are transient in nature, information on corrosion rate, condition of the electrochemical cell, and mechanisms of the corrosion reactions can be obtained rapidly without requiring a system in a steady state. This method has shown promise in laboratory investigations, but is still an exploratory technique, not sufficiently mature for field application.

5.8. Magnetic methods

The main application of magnetic methods is the detection of reinforcing position and size. Pachometers measure cover by detecting variations in an induced magnetic field caused by the ferromagnetic properties of the bars [Bungey 82]. Satisfactory equipment for this purpose is already available. Prototype magnetic instrumentation has also been used for detecting loss of section or fracture of prestressing steel [NCHRPS118 85], but is not generally applicable to conventionally reinforced deck condition assessment.

5.9. Nuclear

The feasibility of using nuclear magnetic resonance (NMR) to determine the moisture content of the concrete and nuclear bombardment to determine chloride content has been investigated, but the methods have not showed promise since they are indirect, expensive, heavy, slow, and NMR requires skilled operators [NCHRPS118 85]. Nuclear bombardment can be used to measure chloride content in the cover concrete by analysis of thermal neutron prompt-gamma and thermal neutron activation-gamma rays. The deck is irradiated with a thermal neutron source. The chlorine nuclides emit a characteristic group of prompt-gamma rays upon absorbing a thermal neutron. The decay of the induced chlorine radioactivity emits activation-gamma rays. Measurement of the prompt and activation gamma rays can be used to find the chloride content profile through the depth of the deck. This method is extremely slow, yielding three to six spot measurements per hour. The capital cost is high and although the FHWA funded development of a prototype instrument, no state highway department has been sufficiently interested to pay fees for demonstration use.

5.10. Infrared thermography

Infrared thermography senses the emission of thermal radiation and produces a visual image from this thermal signal. Thermography, like any system using infrared radiation, measures variation in surface radiance and does not directly measure surface temperature [Joyce 84]. It can be used to detect deterioration that is associated with a thermal anomaly. Surface temperature differences develop due to different rates of heat transfer for sound and unsound areas of deck. A delamination in concrete causes a thermal break so that the delaminated area will absorb and emit heat differently than sound concrete. Debonding between the asphalt overlay and the concrete also causes a thermal break. Thermography can identify this temperature difference between the delaminated or debonded area and sound concrete.

Thermography equipment is available to operate under the following conditions: 70 percent sunshine, a dry deck surface, and wind speed below 15 - 20 miles per hour. There is not a restriction on ambient temperature. This van-mounted equipment can survey a full lane width on each pass, moving at a speed of between 2 and 10 miles per hour. A boom elevates the infrared camera approximately 4.3 m (14 ft) above the deck to reduce distortion due to scanning angle and to increase the field of view. The data is analyzed digitally to produce a map of thermal anomalies. An operator must edit the map against a video recording of the deck to eliminate undesired readings from discoloration, patching, or other surface effects [Kunz 85].

The ability of thermography to detect delaminations on asphalt covered decks is very sensitive to weather conditions such as wind, humidity, and cloud cover [Manning 82]. It is still an open question as to whether or not thermography can reliably discriminate between delaminations and debonding for asphalt-covered decks. Infrared thermography has the advantages of being relatively direct, fast, and high resolution, and the disadvantages of being passive, since it measures re-radiated on solar energy, and sensitive to environmental variables. The potential of this method was judged to be one of the most promising of those surveyed.

5.11. Ground penetrating radar

Radar techniques, developed from U S Army methods for nonmetallic buried mine detection, have been used to sense defects in concrete since the early 1970s. Radar operates by transmitting high frequency electromagnetic waves and sensing the reflections caused by changes in the electromagnetic properties of the material being probed. Two main types of radar systems have been used in highway surveys: continuous wave, swept frequency modulated radar and short pulse, ground penetrating radar (GPR). GPR is the preferred method for highway applications, due to severe speed restrictions on the swept frequency approach since the transmit frequency variation must be slow enough so that the frequency of the return from the top and bottom of the deck are essentially the same [Joyce 84].

Ground penetrating radar (GPR) is a pulsed microwave method that is capable of detecting anomalies associated with a variety of significant physical conditions. GPR is sensitive to: the location and orientation of the reinforcing steel; the concrete cover depth to the top mat of reinforcing steel; the asphalt thickness; the moisture content; the chloride content; and the location and extent of deteriorated concrete. The ability to perform GPR surveys is not weather dependant with the exception of surface moisture on the deck, although temperature needs to be considered when interpreting the results. In a recent test on an asphalt covered deck [Manning 82], radar showed good correlation with known deterioration. There were many false results which showed need for improved signal interpretation. It was judged to offer good potential for a rapid, non-contact, weather independent procedure.

Van-mounted GPR equipment is available [Kunz 85] which operates at 2 to 10 miles per hour and scans a one foot wide strip for each pass. The extent of coverage desired (the spacing between strips) determines the number of passes required to survey the deck. The data is analyzed digitally to identify deck thickness and reinforcing location. Deteriorated areas do produce a distinct signal signature, but additional development of signal processing is required to be able to interpret this data to reliably identify the type, location, and extent of defects. The FHWA is currently developing a van-mounted radar capable of inspecting decks at 40 miles per hour [Joyce 84].

GPR has the advantages of being nondestructive, non-contact, being an active energy method with controllable input, and being highly developed. The collection of data is rapid so the coverage on a deck can be increased. Radar also has the advantage of being potentially capable of determining deterioration related conditions leading to delamination, such as chloride content, moisture content, and rebar cover [Maser 85]. Its main disadvantage is that it is an indirect means of determining mechanical properties and hence can be difficult to interpret. The various investigators cited have proposed different interpretive schemes, but no single scheme has been generally accepted. The volume of data produced is enormous, requiring data reduction and computerized data processing for effective implementation.

6. Comparison of methods

Table 6-I summarizes and compares the methods considered. Sonics, optical, rebound, and penetration tests were dropped from further consideration since they are not applicable to assessment of asphalt overlaid decks. Acoustic emission and x/gamma ray methods do not have appropriate commercially available equipment and are not sufficiently mature as field tests. Sonics are not reliable for overlaid decks and ultrasonics require access to both top and bottom of the deck to obtain meaningful signals. Seismic and microseismic methods are slow, difficult to interpret and require mechanical contact. Magnetic methods are slow, yield a narrow amount of information on deck condition, and already are utilized to an appropriate degree. Nuclear method also yield limited data, are slow, and expensive. Radar and thermography have been identified to be the most promising test methods for further research. Infrared thermography has been more fully developed as a commercial technique of demonstrated value for delamination detection on exposed decks. A better definition is needed of the weather and time of day windows during which thermography can successfully be applied. Ground penetrating radar has been the favored of the two techniques for research potential because it is less sensitive to ambient conditions, because it is equally effective with or without asphalt overlays, and because of the ease of data acquisition and processing. Improved signal interpretation and automation of data analysis are needed to make radar a routine technique.

Table 6-1: Evaluation of test methods

**TEST METHOD	PAVED DECKS	COMMERC AVAIL.	TOP ACCESS	RAPID SPEED	OB-JECTIVE	MOST WEATHER
Hardness	no					
Sonics	no					
Ultrasonics	yes	yes	no			
Microseismic	yes	yes	yes	no	no	yes
Acoust. emission	yes	no				
X and gamma ray	yes	no				
Optical	no					
Magnet.	yes	yes	yes	no	yes	yes
Nuclear	yes	yes	yes	no	yes	yes
Thermography	yes	yes	yes	yes	yes	no
Radar	yes	yes	yes	yes	no	yes

The selection of radar and thermography is in concurrence with the conclusions of other surveys of test methods. Recent studies to identify rapid and more generally applicable methods for bridge deck evaluation have been carried out by the Federal Highway Administration [Joyce 84], the National Cooperative Highway Research Program [NCHRPS118 85], and by the Ontario Ministry of Transport [Manning 82]. These efforts have concluded that among all techniques considered, ground penetrating radar and infrared thermography offer the greatest potential for high speed surveying. These methods are judged to have the most potential for development into routine

**Blanks indicate the method was dropped from further consideration for reasons described in the text.

operational procedures which can be used, in conjunction with existing test practice, to obtain reliable condition assessments of asphalt-covered bridge decks.

However, experiments have not yet been done to establish how these techniques can be applied under field conditions to evaluate in-service decks. Before the promise of these techniques can be realized, a systematic experimental program is required to: demonstrate the range of capabilities of the test methods; evaluate the influence of extraneous environmental variables on the test results; and establish the abilities and limitations of the techniques for determining the state of various bridge decks under a variety of field conditions.

Acknowledgments This paper was made possible by the generous support of the Fannie and John Hertz Foundation and the New England Surface Transportation Infrastructure Consortium.

References

[ASTM 85] American Society for Testing and Materials.
1985 Annual Book of ASTM Standards.
ASTM, 1985.

[Bungey 82] Bungey, J. H.
The Testing of Concrete in Structures.
Surrey University Press (Chapman and Hall, New York), 1982.

[John 81] John, D. G., P. C. Searson, and J. L. Dawson.
Use of AC Impedance Technique in Studies on Steel in Concrete in Immersed Conditions.
British Corrosion Journal 16(No. 2):102-106, 1981.

[Joyce 84] Joyce, R. et. al.
Rapid Non-destructive Delamination Detection.
Technical Report FHWA/RD-84/076, Federal Highway Administration Report, 1984.

[Kunz 85] Kunz, J. T., and J. W. Eales.
Evaluation of Bridge Deck Condition by the Use of Thermal Infrared and Ground-Penetrating Radar.
In *Second International Bridge Conference, Pittsburgh, PA.* June, 1985.

[Manning 82] Manning, D. G. and F. B. Holt.
Detecting Deterioration in Asphalt-Covered Bridge Decks.
Technical Report ME-82-03, Research & Development Branch Ontario Ministry of Transportation & Communications, September, 1982.

[Manning 86] Manning, D. G., C. J. Arnold, K. C. Clear, and R. A. Dorton.
Technical Research Area #4 Detailed Planning for Research on Bridge Component Protection.
Technical Report NCHRP Project Number 20-20, Pre-Implementation Activities for the Strategic Highway Research Program (SHRP), January, 1986.

[Maser 85] Maser, K. R.
Intelligent Systems for the In-Situ Evaluation of Materials and Structures.
Technical Report Final Report NSF/SBIR CEE 8460816, National Science Foundation, July, 85.

[Murphy 85] Sarnes, F. W., V. J. Murphy, and F. C. Sankey.
Integration of Geophysical Techniques and Direct Inspection for Bridge Deck Evaluation: Market Street Bridge.
In *Second International Bridge Conference, Pittsburgh, PA.* June, 1985.

[NCHRPS118 85] National Cooperative Highway Research Program.
Detecting Defects and Deterioration in Highway Structures.
Technical Report NCHRP Synthesis 118, Transportation Research Board, 1985.

[NCHRPS57 79] National Cooperative Highway Research Program.
Durability of Concrete Bridge Decks.
Technical Report NCHRP Synthesis 57, Transportation Research Board, 1979.

[Park 80] Park, S. H.
Bridge Inspection and Structural Analysis: Handbook of Bridge Inspection.
S. H. Park, Trenton, N.J., 1980.

68th Street Intake Crib-Inspection and Repair

Michael J. Garlich, S.E., M. ASCE*

Abstract

After nearly 90 years of exposure to the ravages of Lake Michigan weather, the 68th Street Crib water intake structure suffered partial failure of part of its protective breakwater. A detailed structural inspection both above and below water was undertaken to determine the extent of damage and potential for further deterioration.

As a result of the inspection, a two step approach to repairs was adopted. The first stage involved reconstruction of the failed timber crib breakwater as a stone revetment. The second stage involved design of a tied back sheet pile wall along the outer breakwater wall. Since rework to the breakwater concrete cap was required for any repair scheme, the new concrete was designed to act as a ring beam resisting the tie back forces.

Introduction

The City of Chicago, as well as numerous surrounding communities, are supplied water from Lake Michigan which is drawn in through intake structures located out in the lake. One of these, the 68th Street Crib and breakwater was built in 1893-1896 as an intake structure supplying water through a three-mile long tunnel to the 68th Street Pumping Station. The crib is no longer in service as an intake structure, but it provides year-around living quarters for the tenders of the adjacent newer Edward F. Dunne Crib which is connected to the 68th Street Crib by a steel bridge. Figures 1 and 2 show both crib structures and the breakwater.

The 68th Street Crib consists of a stone masonry superstructure supported on stone-filled timber cribs sunk into the lake bottom. The top portion of the crib is filled with concrete-filled bags and cast-in-place concrete. In the 1930's, some additions were made to expand the living quarters of the crib.

*Vice President, Collins Engineers Inc., 600 W. Jackson Blvd., Chicago, IL 60606

Figure 1. View of 68th Street Crib, Breakwater and Edmund Dunn Crib looking from the west

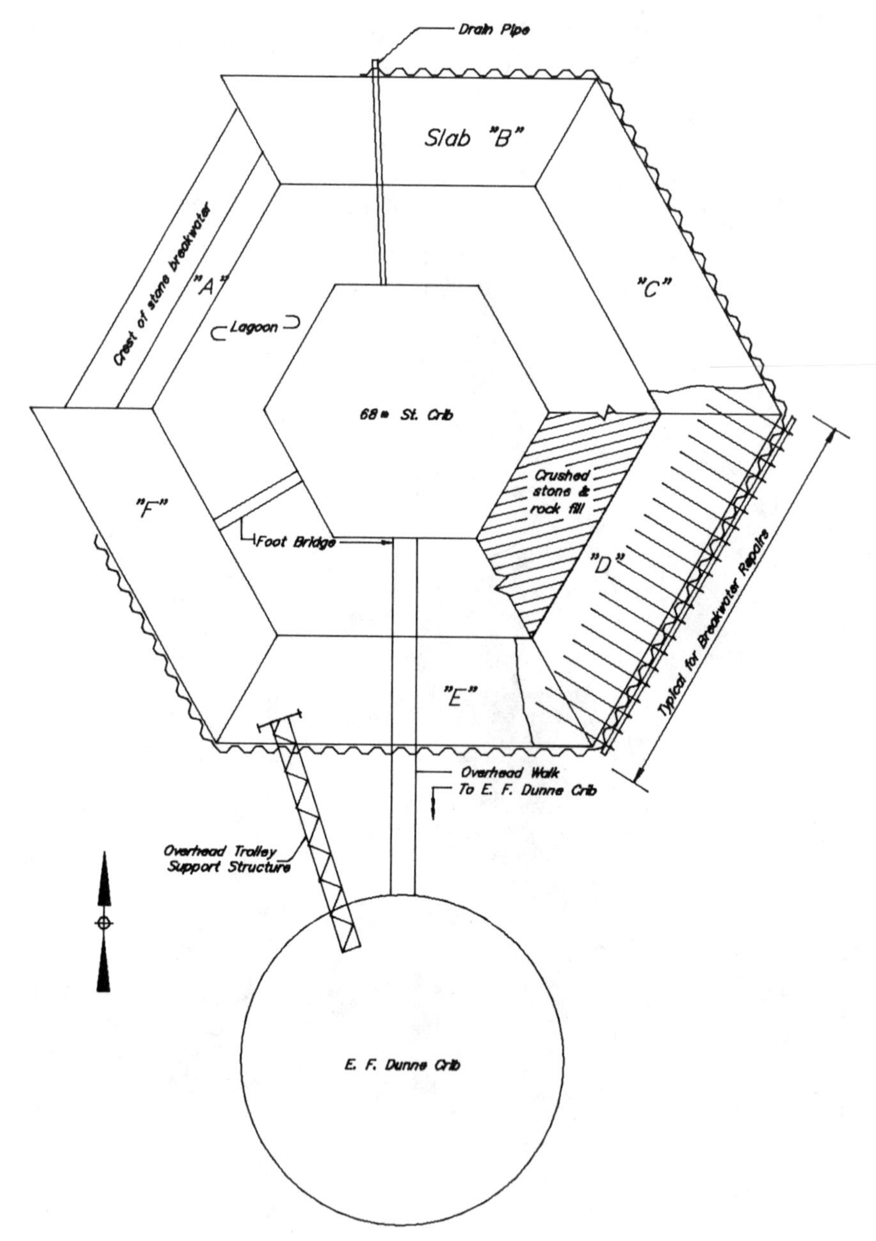

Figure 2. General Plan

A hexagonal breakwater, 120 feet (36.6m) on a side, surrounds the 68th Street Crib. The breakwater is of stone-filled timber crib construction with a concrete cap. The timber cribs were constructed in sections and floated to the site where they were sunk in place by filling them with stone. Such structures are common in the Great Lakes. For the 68th Street cribs, oak was used for the walls and cross ties from the crib top to 9 feet (2.25m) below the waterline, and pine or hemlock was used from that point to the lake bottom. Members were connected utilizing half lap dovetail joints and drift pins. The gaps between the assembled sections were also filled with stone. A cast-in-place concrete cap forms the top surface of the breakwater. Water depth at the crib is approximately 32 feet (9.75m).

Method of Investigation

In June, 1983, a detailed underwater inspection was made of the 68th Street Crib and the surrounding breakwater from the area near the waterline to the lake bottom. A visual inspection was made of the exterior elements of these structures and accessible interior passages and voids. The inspection included probing and sounding in areas of apparent distress.

Specifically, the areas inspected included the exterior surface of the breakwater; the inside of void areas; the spaces between adjacent timber crib sections; the interior, or lagoon surface of the breakwater; the interior of intake ports through the breakwater; the exterior surface of the 68th Street Crib. The divers attempted to inspect inside four intake ports of the crib. They were able to swim through two of the ports to the interior of the crib, but silt accumulation prevented entrance through two others.

The underwater inspection was conducted by a registered structural engineer-diver, assisted by an engineer-diver and a technician-diver. The divers generally used scuba diving equipment for mobility, but used surface supplied lightweight diving apparatus with communication equipment in areas which required copious notes. Findings were reported to a tender-notetaker on the surface. The divers worked from the surface of the breakwater. In general, visibility was relatively good. Light penetrated to the lake bottom and visibility ranged from about six to ten feet (2 to 3m). Inside the intake ports the divers used underwater lights. Underwater 35mm cameras and strobes were used to document the inspection findings.

Soundings were made of the lake bottom along the inside and outside of the breakwater and along lines parallel to and at distances of 25 feet (7.6m) and 50 feet (15.2m) from the outside face of the breakwater. The soundings along the faces of the breakwater were made using a lead line and the soundings at 25 feet (7.6m) and 50 feet (15.2m) from the breakwater were made using a continuous recording type fathometer.

The above water portions of the breakwater were examined for deterioration also, and a survey of elevations was made.

Existing Conditions

In general, the crib substructure was found to be in good condition. There were, however, a few areas of deterioration of stone, concrete and timber near the top of the timber-filled crib. These defects were not significantly reducing the strength or stability of the structure but provided an area for more significant distress to develop unless protective measures were taken. The exterior masonry portion of the crib building is carried on the exterior timbers of the stone filled crib. Gaps existed at the top of the crib that reduced the support for the masonry walls. On the south side of the crib building there were some minor cracks in the lower two courses of stone above the waterline. These cracks were above a void area at the top of the timber crib. The floor of the crib building is supported on concrete-filled bags and the stone fill within the timber cribs. This slab is quite thick and there was no evidence of distress caused by the voids at the top of the timber-filled crib. It could be expected, however, that if the voids were not corrected, the loss of stone and concrete bag fill would continue slowly and eventually cause distress to the structure.

The breakwater had suffered significant distress along the west side, which seriously weakened it and reduced its ability to protect the crib building. Approximately half of the east side of the west breakwater had failed. If corrective action were not taken, it could be expected that the damage would extend to adjacent sections and become progressively worse. The east side of the breakwater showed evidence that it was also starting to fail in a manner similar to the west side. It could not be determined with certainty if this condition was progressing, but it did appear that the conditions necessary for a local failure were present. The condition of the inside of the breakwater on the south side of the crib was not as severe as the other conditions, but local failures of the timber were evident. It was also not known if this condition was progressively deteriorating, but the conditions antecedent to failure appeared present. Several local areas of failed timber were found in other sections. In

Figure 3. Deterioration of top of breakwater

Figure 4. Deteriorated half lap dovetail connection of crib through timber to face timber

addition, stone had fallen out of the crib sections at the corners of the breakwater to various depths.

The concrete cap on the breakwater was in fair condition except for a section on the west side approximately 60 feet (18.3m) long which had settled about 5 feet (1.5m) and cracked, and areas of surface spalling extending about 6 inches (150mm) deep. Minor differential settlement had also occurred all around the breakwater.

Soundings of the lake bottom adjacent to the breakwater indicated that stone rubble from the failed section of breakwater extended outward from the exterior wall about 30 feet (9.2m). Some additional spillage was also noted by soundings, particularly around corners. Otherwise, the water depth was found to range from 28 to 31 feet (8.5 to 9.5m). No evidence of significant scour was found and riprap protection was found in place along all outside faces of the breakwater.

Recommendations

An evaluation of the existing conditions and their future ramifications as well as budget considerations resulted in the recommendation of a two step repair process.

Immediate repairs to the failed west breakwater were recommended to prevent further rapid deterioration of this and adjacent areas. Emergency funds were available for this work. This also would serve to minimize the potential need for emergency repairs during winter months when construction costs in Lake Michigan are extremely high.

It was also recommended that repairs to the remaining areas of breakwater and the crib be undertaken as soon as monies could be budgeted for their construction. The City agreed with this approach to repairs, and design of the west breakwater reconstruction was started immediately so construction could be completed prior to the onset of winter weather.

First Stage Repairs

Several design options for stabilizing the failed west breakwater were considered. Construction of a sheet pile wall with a new concrete cap was ruled out because of the large amount of spilled stone from the failed section. Piling could not be driven through the stone, and removal of the stone to facilitate driving would have caused further deterioration of the structure.

Consideration was also given to restoring the lake side face of the breakwater with cast-in-place concrete. This would be placed atop the existing spilled stone and tied back to the remaining sound sections of the structure. Some underwater placement of forms, tie steel and concrete could be required. A similar repair scheme had been proposed for the Two Mile Crib in 1918.

The scheme finally selected was to construct a stone armored revetment against the failed outer face of the breakwater. Sizing of the stone was based on a 10.8 foot (3.3m) wave height with a period of 7.3 seconds. Some overtopping of the crest was considered acceptable because of the presence of the lagoon between the breakwater and the crib. A cross section through the revetment is shown in Figure 5. In constructing the revetment, a filter cloth was first placed on the lake bottom extending up a minimum of five feet onto the spilled stone. A layer of 2 to 4 ton (1800 kg to 3600 kg) stone was then placed to a thickness of approximately six feet (2m) topped with an eight foot (2.6m) thick layer of 5 to 8 ton (4500kg to 7600kg) armor stone. Stone was supplied from quarries approved by the Corps of Engineers.

Construction of the concrete cap was delayed to the second phase of repairs to allow for settling and "working in" of the stone to take place. No particular problems were encountered during construction though some experimentation was required by the contractor to develop an efficient technique for placing the filter fabric. Repairs were successfully completed before the onset of winter weather. Figure 1 shows the crib and completed stone revetment.

Second Stage Repairs

The second stage of repairs was desgined to stabilize the remaining sections of breakwater and the crib. The scheme chosen was significantly influenced by the following considerations:

- Since the crib structure no longer serves as a water intake, the horizontal intake ports originally provided in the breakwater and crib structure could be closed. Water level within the lagoon would still fluctuate with lake levels by infiltration through the riprap breakwater.

- The crib tenders use the breakwater for exercise and relaxation in warm weather. Thus repairs to produce a smooth level surface to the breakwater cap concrete were desirable.

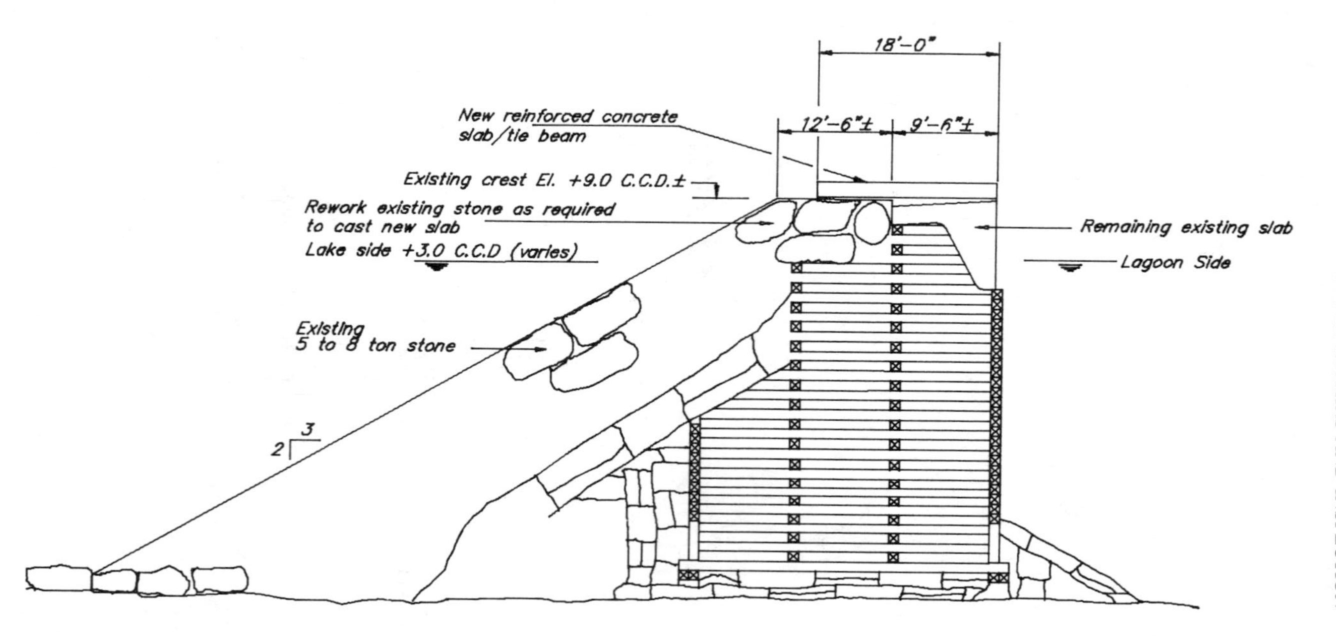

Figure 5. Section through stone revetment repair to breakwater

. The recent rise in levels of Lake Michigan - projected to continue for at least several years - has caused increased overtopping of the breakwater.

. Supply tugs visit the crib several times each week. Current high lake levels make docking difficult. Thus, a higher breakwater was desirable. It also meant that any breakwater repair scheme had to leave room at the southwest and southeast breakwater sections to tie up a 70 foot (21.3m) tug, and that the tug be able to negotiate the space between the 68th Street and Edwin Dunn cribs to off load heavy items by the overhead bridge crane.

Grouting the breakwater stone fill to form a solid "concrete" mass was considered impractical due to the difficulty of sealing the outer faces to contain the grout. Also, the cost required to drill through the rubble stone and then fill the large void volume present in the large rubble stone used for crib fill was very high.

Spilled stone and 90 years of accumulated debris are present in the lagoon area. Driving piling through this debris would be difficult and the configuration of the lagoon also make debris removal costly. Thus, use of double sheet pile walls for breakwater reconstruction was dicarded.

The requirement for boat movement and docking along the breakwater restricted the areas where a stone revetment could be constructed as it was for the first phase repair. Use of a revetment in conjunction with some sheet pile dock areas was also considered, but proved costly.

The solution chosen was possible only because the crib itself no longer is used to draw in water, since it effectively closes off the various intake ports. This repair scheme is shown in Figures 2 and 6 and has the following major points:

. A sheet pile wall is used to stabilize and protect the outer face of the breakwater. Thus, tug tie-up and maneuvering space is maximized. Dredging of some spilled material will be needed prior to driving sheeting. Sheeting is placed two feet outside the existing breakwater to lessen dredging effects and miss displaced timbers.

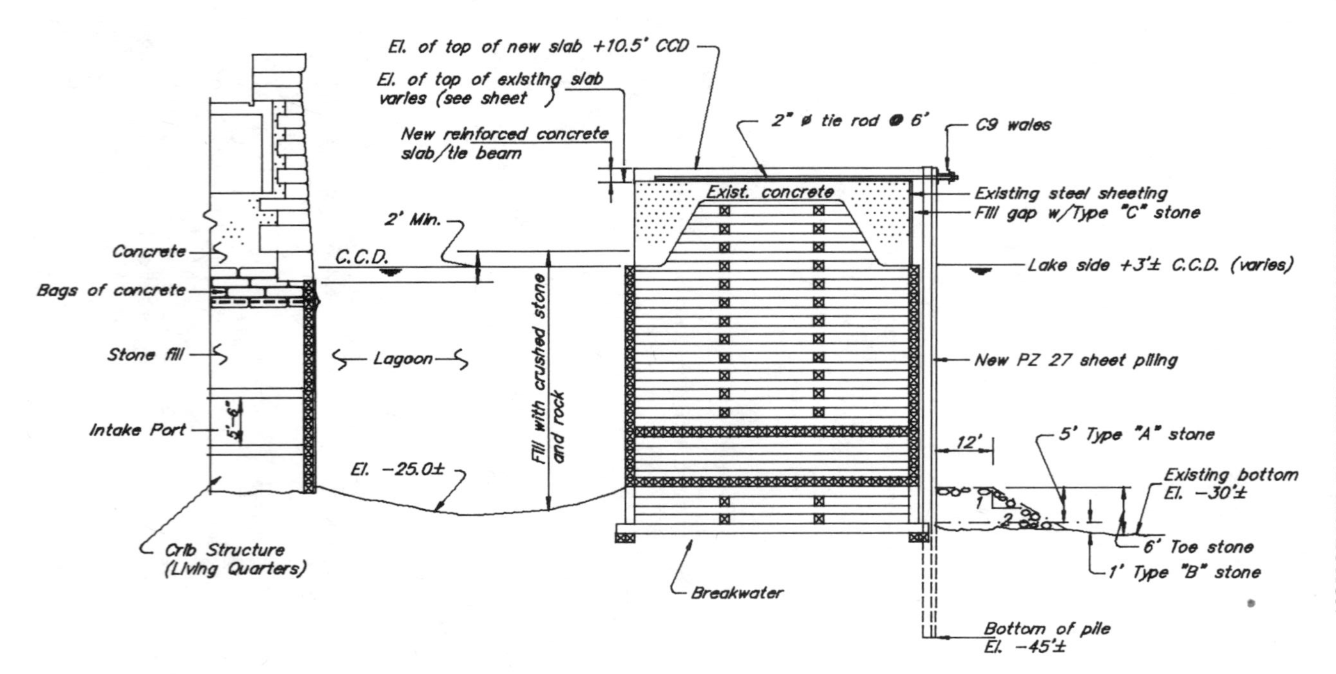

Figure 6. Section showing sheet pile repairs to breakwater

- The breakwater is topped with a reinforced slab. Serving as a tension ring for anchorage of the sheet pile tie rods, the slab also yields a higher breakwater and a safe top surface.

- The inside of the breakwater and outer face of the crib is stabilized by filling the annulus or lagoon area with stone.

Configurations incorporating heavy sheet pile and structural shapes into cantilever walls were studied, but costs were much higher than the tie back system. Use of readily available material was also desired which the chosen scheme satisfied. The considerable economy gained was primarily due to the need for a new breakwater cap for all schemes. Thus the concrete cap serves double duty, producing a structure similar to those used for cofferdams or cells. The slab was designed considering behavior as a continuous ring beam subject to bending and shear as well as a tension ring.

Sheeting and tie rod design considered both active and at rest conditions within the deteriorated breakwater stone fill. The lake bottom in the area is medium dense, fine to medium sand. Toe stone was provided to prevent undermining and also reduce the sheeting stresses. Use of random stone fill within the annulus provides an economical means of supporting both the inside of the breakwater and outside of the crib.

A new timber fender system to accomodate rising, and future falling, lake levels was incorporated into the repairs along with several other minor items. As part of the final in-house design review, comments were solicited from local marine contractors on constructability.

Conclusion

That the 68th Street Crib survived nearly 90 years of service in Lake Michigan weather in good condition is a testament to the knowledge and skill of our predecessors. Though its function has changed with time, and this change contributed to the choice of repair plans, it remains a valuable part of Chicago's water supply system. The completed repairs will enable it to continue to serve its present functions.

AUTHOR INDEX

Page number refers to first page of paper.

SUBJECT INDEX

Page number refers to first page of paper.